Artificial Intelligence and National Security

Reza Montasari
Editor

Artificial Intelligence and National Security

Editor
Reza Montasari
School of Social Sciences, Department of
Criminology, Sociology and Social Policy
Swansea University
Swansea, UK

ISBN 978-3-031-06708-2 ISBN 978-3-031-06709-9 (eBook)
https://doi.org/10.1007/978-3-031-06709-9

This Springer imprint is published by the registered company Springer Nature Switzerland AG
The registered company address is: Gewerbestrasse 11, 6330 Cham, Switzerland

Dedicated to my wife, Anna, without whose unfailing love and support I could not have completed this work.

Contents

Artificial Intelligence and the Spread of Mis- and Disinformation

Annie Benzie and Reza Montasari

Abstract In a so-called post-truth era, research on the subject of the spread of mis- and disinformation is being widely explored across academic disciplines in order to further understand the phenomenon of how information is disseminated by not only humans but also the technology humans have created (Tandoc, Sociol Compass 13(9), 2019). As technology advances rapidly, it is more important than ever to reflect on the effects of the spread of both mis- and disinformation on individuals and wider society, as well as how the impacts can be mitigated to create a more secure online environment. This chapter aims to analyse the current literature surrounding the topic of artificial intelligence (AI) and the spread of mis- and disinformation, beginning with a look through the lens of the meaning of these terms, as well as the meaning of truth in a post-truth world. In particular, the use of software robots (bots) online is discussed to demonstrate the manipulation of information and common malicious intent beneath the surface of everyday technologies. Moreover, this chapter discusses why social media platforms are an ideal breeding ground for malicious technologies, the strategies employed by both human users and bots to further the spread of falsehoods within their own networks, and how human users further the reach of mis- and disinformation. It is hoped that the overview of both the threats caused by and the solutions achievable by AI technology and human users alike will further highlight the requirement for more progress in the area at a time when the spread of falsehoods online continues to be a source of deep concern for many. This chapter also calls into question the use of AI to combat issues arising from the use of advanced Machine Learning (ML) methods. Furthermore, this chapter offers a set of recommendations to help mitigate the risks, seeking to

---

A. Benzie (✉)
Hillary Rodham Clinton School of Law, Swansea University, Swansea, UK
e-mail: A.R.Benzie@Swansea.ac.uk

R. Montasari
School of Social Sciences, Department of Criminology, Sociology and Social Policy, Swansea University, Swansea, UK
e-mail: Reza.Montasari@Swansea.ac.uk
URL: http://www.swansea.ac.uk

R. Montasari (ed.), *Artificial Intelligence and National Security*,
https://doi.org/10.1007/978-3-031-06709-9_1

explore the role technology plays in a wider scenario in which ethical foundations of communities and democracies are increasingly being threatened.

Keywords Artificial intelligence · Misinformation · Disinformation · Social media · National security · Cyber threats · ML · Software robots

1 Introduction

The way that people communicate with each other has changed (Bakardjieva, 2005). Social media platforms such as Facebook and Twitter allow people to instantly access information on a wealth of subjects from users anywhere in the world, in real time. The World Wide Web which had its roots firmly in creating an environment which fostered freedoms (Benedek & Kettemann, 2020) and which connected people globally has replaced the millions of offline communities separated by distance, language, and culture, with a single global community. One where each of the billions of individuals who have access to an internet connection can contribute to the creation and dissemination of information quickly, and at low cost (Kreps, 2020). In reality, the abundance of information in a globalising world is not a new cause for concern. From James Madison in the nineteenth century to the work of Habermas and the many researchers who have followed since, there is one constant: a fear of both the controls and lack thereof, surrounding the ever-growing volume of information (Habermas, 1984). Offsetting the benefits of a more connected world are a string of worries as far-reaching platforms have become breeding grounds for mis- and disinformation (Tandoc Jr, 2019), which is disseminated by both humans and machines often with malicious underlying intentions. As such, the subject of how falsehoods are spread online has rightfully become a topic of intense interest for academics across an array of disciplines, such as computer science, psychology, and politics (Shu et al., 2020). This study was carried out with the aim of highlighting literature integral to this subject. The study adds to the existing body of research offering further insight into not only the manner in which mis- and disinformation is spread but also the challenges this presents. The study also focuses on ways in which the risks associated with widespread access to falsehoods may be mitigated. It is perhaps vital that we consider the challenges that modern technology has brought into the equation, in particular AI.

The remainder of this chapter is structured as follows. In Sect. 2, the meaning of information and truth in a post-truth world are explored. A particular focus is placed on the fine line between lies which refer to truth, and post-truth which seeks to delegitimise truth completely. The definitions of essential terms such as ‘misinformation’ and ‘disinformation’ are discussed, including exploration around the question of intent which distinguishes the two terms by definition. In Sect. 3, the role of AI in the spread of mis- and disinformation is considered, focusing on the types of disinformation and how fabricated content using techniques such as Generative Adversarial Networks (GANS), deepfakes, and multimodal content may be detected. This includes multifaceted approaches focusing on the content, organisation, emotion, and manipulation pertinent to the media content. In this section, the role of social media and AI bots are discussed, showing that although users may not actively consume content, they might disseminate said information, thus creating clusters within the network which align with and confine users to their personal beliefs. Section 4 further considers both the micro and macro impacts the spread of misinformation has on individuals and society as a whole. This includes a discussion on whether it is possible for disinformation to sway public opinion on such a scale as to conflict with democratic systems. Finally, ways in which the impact of widespread disinformation may be mitigated are considered, such as the introduction of a mass-collaboration model, increased education and focus on media literacy for users, as well as increasing the burden of responsibility of social media companies to find and employ effective solutions. Finally, after discussing recommendations for how to mitigate the risks of spreading falsehoods online, the synthesis of literature explored is concluded by providing final thoughts and considerations for the future, particularly in terms of research which would be of further benefit to the academic community.

2 Misinformation Versus Disinformation

2.1 Information and the Meaning of Truth in a Post-Truth World

The conversation around information is one which has been ongoing for several decades. It is often argued that untrue information does not qualify as information (Floridi, 2004). In his 'Outline of a Theory of Strongly Semantic Information', Floridi presents the argument that in order to qualify as information, the sentence must be true. Under this theory, many statements which although may be considered important data to some, could not be considered information, if it is the case that there is insufficient evidence to suggest the statement to be concretely true (Floridi, 2004). Countering this theory, Fetzer (2004) suggests that even the mere existence of the terms 'misinformation' and 'disinformation' indicates that false information is not anomalous (Fetzer, 2004). Similarly, a conclusion drawn by Buckland argues that it is complex to definitively state that anything is not informative in some way; therefore, it must be defined as being information. With this in mind, a statement which is deemed as informative could therefore be considered as information, albeit false or misleading (Buckland, 1991). Outside of one's own experiences, our perception of what is true is based on representatives of the truth that we deem reputable, for example, the news. According to Mingers (2001) and Carsten Stahl (2006) 'Such meaning is only relevant if information can affect actions or perceptions' (Carsten Stahl, 2006; Mingers, 2001). It is perhaps also key to consider the distinction between having information and being informed. While being informed requires that one has information, it is a much grander condition than having access to masses of information (Webster, 2014). In his 2006 work 'The Logic of Being Informed', Floridi considers that 'the informativeness of a message – raises issues of e.g., novelty, reliability of the source and background information' (Floridi, 2006).

As such, being informed is therefore dependent on the relative informativeness of the information. Furthermore, given that an overload of information, some of which will derive from unreliable sources, more information does not necessarily equate to being well-informed. If we understand that information does not equate to truth, then we can also establish that truth can be regarded as an 'ideal' type of information whereby the information provided in the statement is true (Carsten Stahl, 2006). This is echoed by Habermas (1984), who argues that ultimately rational speakers deem the truth to be the ultimate goal of communication (Habermas, 1984). However, there are specific terms used to represent different types of falsehoods. Pertinent to this chapter are, of course, the terms 'misinformation' and 'disinformation'. Another important consideration is the misalignment between the concepts of lies and post-truth. Bufacchi (2021) argues that while lies still predominantly refer to the truth, post-truth seeks to destroy the entire infrastructure on which the truth is based. 'Post-truthers are different: their aim is to delegitimize truth, since this is the best way to disarm the threat truth poses to them' (Bufacchi, 2021). The article also argues that although post-truth is a popular subject in the modern world, it is far from a new concept. As is outlined in the 1967 essay 'Truth and Politics' (Arendt, 1967), politics and truth are not necessarily a natural combination. Fifty-five years later, this concept has not changed. However, with platforms on social media being readily used, it is much easier for individuals to spread their ideologies and essentially employ modern technologies to act as a stage from which to shout. With new data being created and uploaded in real time, hunting for the truth could perhaps draw comparisons with hunting for a needle in a haystack, if the needle were to resemble every piece of hay in the stack. How does one distinguish truth from falsehoods? How can information be sorted into piles of information, and mis- or disinformation? Approaches to how this has been achieved using ML algorithms will be drawn upon in Sect. 3.

2.2 *The Differences Between Misinformation and Disinformation*

Misinformation may be defined as 'a claim that contradicts or distorts common understandings of verifiable facts' (Guess & Lyons, 2020). It has also more generally been defined as 'false or misleading information' (Lazer et al., 2018). While misinformation is false information which may be disseminated mistakenly, the OED defines disinformation as 'the dissemination of deliberately false information' (European Parliament, 2015). Guess and Lyons (2020) define the difference between the two in the intent (Guess & Lyons, 2020). While misinformation may be unintentionally false, disinformation is purposefully created to deceive or mislead (Ciampaglia et al., 2018). According to Habermas (2003), misinformation is not perceived to be an issue as it can be analysed and critiqued within discourse, and the speaker must explain their reasoning (Habermas, 2003). This is different to disinformation which can be considered more dangerous in the sense that it is deliberately disempowering, and the speaker therefore disregards the individual who may be alienated. This communicative approach can therefore be seen as strategic (Habermas, 2003). In fact, according to Vasu et al., disinformation is 'the most onerous given its impact on national security and social cohesion' (Vasu et al., 2018). In this article, the authors present the idea that falsehoods for entertainment may also be dangerous as some believe parody to be true. The issue pertaining to this is that extremist views may be masqueraded under the guise of something else using irony to make the idea seem mainstream. Essentially, this strategy allows individuals to push their ideas on a public stage, but gives them enough leeway to retreat if met with resistance (Vasu et al., 2018). According to his work 'Disinformation: The Use of False Information', disinformation bears similarities to lying, as we can state that lies are pieces of information 'that are false, that are known to be false, and that are asserted with the intention to mislead, deceive, or confuse' (Fetzer, 2004). The article draws upon a theory.

1. Offering information when not in the position to provide such knowledge.
2. The ignoring or bypassing of relevant information which may alter the opinion, or argument.
3. Attacking someone, such as an author, on 'misleading grounds' which are irrelevant to the author.
4. Creating 'a biased impression of a specific study by simply excluding the most significant features' (Fetzer, 2004).
5. Deliberately misleading someone by only using evidence which supports the preferred argument.

Asking how the spread of misinformation threatens individuals bears similarities with questioning the harms of largescale deception. As an example, a study in 2020 reviewed the perceived credibility of various AI models in the face of shaping public perception on foreign policy (Kreps et al., 2020). The results demonstrated that people are unlikely to be able to differentiate between AI and human-generated text, thus proposing that there indeed exists a 'propensity for manipulation' in terms of users of the online space (Kreps et al., 2020). The article suggests that it is not the case that the content produced by AIs are able to outright change one's opinion or political beliefs, but rather they ignite a sense of confusion or distrust in the individual. In other words, it sows the seeds which may later grow (Weedon et al., 2017). Norris (1996) expresses the threat of people believing falsehoods as a way in which common reference points within society are broken down, thereby undermining systems (Norris, 1996).

3 AI, Bots, and the Spread of Mis- and Disinformation

3.1 Definitions of AI and ML

According to 'Aspects of AI', AI mimics activities which are deemed to be human-like, such as making rational decisions to solve problems. It is used in most sectors today and is very much embedded in our everyday lives (Neelam, 2022). AI was introduced in the 1950s by Alan Turing in his notable work 'Computing Machinery and Intelligence' which presented the idea of machines behaving with intelligence (Turing, 1950). Developing this theory more recently, Russel and Norvig conceptualised the idea of machines that can plan ahead and act on these plans independently (Russell & Norvig, 1995). According to El Naqa and Murphy, ML is a branch of AI which involves computational algorithms that are capable of learning from their environment and, therefore, imitating human behaviour (El Naqa & Murphy, 2015). In 1983, it was thought that ML had three research focuses: task orientated studies; cognitive simulation; and theoretical analysis. In his chapter 'Why Should Machines Learn?', Simon theorises that the main distinction between the learning process of humans and machines is that it generally takes far less time for a machine to learn how to perform a task, and how to do so to an extremely high standard (Simon, 1983). Consider the task of playing a game. We have seen that machines are not only able to learn the rules of a game immediately, but their ability to plan ahead means that they can improve their performance quickly to the point of becoming an expert in mere hours (Simon, 1983). However, ML is not restricted to gaming environments, and is commonly employed in a number of malicious settings, including in the fabrication of online content, as is discussed below.

3.2 Types of Disinformation and Detecting Fabricated Content

3.2.1 GAN Images

In recent years, it has been common to come across fabricated images which may be widely circulated on social media platforms. Generative Adversarial Networks (GANS) is a method in which fabricated images are machine-generated, often being realistic enough to manipulate users (Gragnaniello et al., 2021). GAN may be used to restore images and brings to the table some exciting features. However, enhanced photorealism has made the task of distinguishing false media content from authentic content a highly complex one (Gragnaniello et al., 2021). It is widely agreed that there is an urgent need for automatic tools which can detect synthetic media as often to the naked eye, false media is undetectable and often goes undiscovered (Shu et al., 2020). Over time, the presence of anomalies in images has diminished, including dissymmetry. However, studies have shown that invisible 'artificial fingerprints'

(Shu et al., 2020) can be recovered from a GAN-generated image which, as evidence, is much more difficult to eradicate completely. A number of studies focusing on GAN detection focus on training a classifier on a dataset of GAN-generated images produced by a pre-trained GAN model. The study, 'Detecting and Simulating Artifacts on GAN Fake Images' takes a slightly different approach by taking a closer look at the GAN generation pipeline, particularly at up-sampling layers which help to increase the resolution of the image (Zhang et al., 2019). This is before training the detection classifier with the frequency spectrum and not the raw RGB pixel. The results of this study demonstrate that this significantly increases the ability of the classifier to generalise (Zhang et al., 2019).

3.2.2 False Videos and Deepfakes

Technological advancements have made it possible to use generative adversarial networks to replace faces of a person in video content with the face of another person entirely. As a result, this creates content which may successfully manipulate viewers into believing that the content is authentic. Research into the impact of so-called deepfakes, as well as deepfake detection, has increased in recent years. George (2019) found that in cases where deepfakes are used to undermine politicians and political campaigns, this has the potential to have a resulting impact on the public sentiment towards the politician or political party in question (George, 2019). In terms of deepfake detection, interesting research strategies have been investigated. One such strategy is by Li et al. (2018), who sought to detect abnormalities in blinking under the assumption that this movement may appear stifled or may not appear at all in an artificial video (Li et al., 2018; Yuezun et al., 2018). This compares to studies such as that carried out by Güera & Delp (2018), who used convolutional LSTM to detect discrepancies between video frames in the facial features, for example, abnormalities in the lighting (Güera & Delp, 2018).

3.2.3 Multimodal Content

While most disinformation is based in one singular modality such as text or video, multimodal strategies are often used to make news articles more believable by having accompanying text alongside a fake video, for example. According to (Singh et al., 2020), it is essential to consider the authenticity of the content as a whole instead of each individual component, particularly given that news articles are becoming increasingly multimodal in structure. The detection of multimodal content has predominantly focused on intelligent text-processing strategies. The study of Singh et al. (2020) presents the need for a multimodal detection framework which has the capacity to review and identify abnormalities in different areas of the news item. They therefore consider a multifaceted approach to the detection of disinformation using a combination of text-based and visual features, and their relationship with the following four categories:

- Content: the topics that are identified in the item.
- Organisation: how these topics are presented to the reader or viewer.
- Emotion: this may include facial expressions present within images or video content.
- Manipulation: how the information may be distorted in favour of a particular viewpoint. The level of manipulation present in the text may be measured by excessive use of second-person pronouns (Horne & Adali, 2017), or the number of features within an image that have been tampered with.

3.3 Software Robots

Software robots, or bots for short, may be created for various purposes depending on the intent of the creator. This may be simply to automate a process on a website, such as a chatbot which directs the user to relevant documentation to help resolve a query, or to create and interact with content online. Bots may require assistance to function by a human or be completely automated. Bots often do not work alone, and a group of bots working collaboratively is known as a botnet (Burkhardt, 2017). Bots are programmed algorithms which, in this case, assist with the dissemination of information by directing web content such as articles to the newsfeeds of social media users. This is a tactic which is used by the likes of political parties and news agencies which is extremely powerful. According to Burkhardt, bots are often programmed to identify the types of content the user tends to click on and push additional content which is similar to this. In this way, a user may only be able to see articles which are already aligned with their own views or published by similar sources (Burkhardt, 2017). This information may also be directed to those within the user's network, creating confined bubbles of information. Burkhardt argues that individuals within these echo chambers may feel as though their ideologies reflect those of the majority of the population, as they no longer see information that conflicts with their opinions (Burkhardt, 2017). The benefit of bots for parties wishing to disseminate information is that they work around the clock and work much faster than humans could. In the last few years, bots have increasingly been designed to spread not just targeted information, but misinformation. What is more, humans assist bots with their duties by engaging with the content by sharing across their own networks. In fact, one study showed that 55% of users spent less than 15 seconds on a page (Pedro Baptista & Gradim, 2021). Effectively, information is disseminated by bots and users who have often not read the content. Some users would not question this content, as they prefer to consume information that is within their own system of beliefs.

3.3.1 Types of Bots

According to Himelein-Wachowiak et al., content polluters are one type of bot which seeks to 'disseminate malware and unsolicited content' (Himelein-Wachowiak et al., 2021).

Spambots Giorgi et al. (2021) present their work which discusses the increased online presence of social spambots, a class of bot which can be defined as 'a computer algorithm that automatically produces content and interacts with humans on social media, trying to emulate and possibly alter their behavior' (Giorgi et al., 2021; Ferrara et al., 2016).

Social bots Social bots are programmed to engage with content online by interacting with posts from users as well as posting content themselves, acting in the same way as human users. Studies such as those by Shao et al. (2018) present that social bots are partly responsible for the infodemic, as bots target users who may be vulnerable to manipulation and thus share content shared by the bots to their own network (Shao et al., 2018). The main difference between spambots and social bots is that social bots act in a similar fashion to human users tend to act online, whereas spambots do not wish to hide their bot-like features.

Cyborgs Cyborgs are also common on social media platforms and have been defined as 'either bot-assisted humans or human-assisted bots' (Zhang & Ghorbani, 2020). For example, a user may set up automated posts and participate every so often by engaging with other users in their network. By demonstrating both manual and automated behaviour, lines may be increasingly blurred, which increases the difficulty for a user when it comes to differentiating a bot from a human user. A study by Chu et al. (Chu et al., 2012) aimed to classify Twitter accounts into human, bot, and cyborg categories with the ultimate goal of helping users to identify and avoid malicious users online and thus increase safety online. As such, the differences are divided into three categories: behaviour, content, and account properties, before data analysis is conducted on a dataset of over 40 million tweets (Chu et al., 2012). In terms of user behaviour, this included retrieving insight into whether tweets were posted periodically or sporadically, as posting according to a timetable is often a sign of automation. The study also expressed the need to check if tweets contained spam using text patterns and the account properties of the user (Chu et al., 2012).

3.4 Social Media and the Spread of Mis- and Disinformation

Many bots do not exist to participate in malicious activity, but instead actively help users navigate information online, such as chatbots which aid users to access information 24/7 (Chen et al., 2017). Despite this, as can perhaps be demonstrated by the above studies, research surrounding bots often depicts social media as an ideal location for bot detection. Chen et al. (2017) state that approximately 9–15% of active Twitter accounts are bots. Himelein-Wachowiak et al. (2021), refer to the burden of conflicting information being provided online surrounding the Covid-19 virus as an 'infodemic' (Himelein-Wachowiak et al., 2021). It presented that 33% of the US population reported seeing misleading information regarding the pandemic on several occasions on social media platforms. The article noted that within a subset of tweet data, 25% were deemed to contain misinformation; however, there was reportedly no difference between engagement levels of tweets containing misinformation or those containing accurate and reputable information (Himelein-Wachowiak et al., 2021). This emphasises that once information has been posted online, regardless of accuracy, there is the same level of interaction with the information by human users. At their core, social networking platforms represent networks of trust, which allow users to share information to their communities of friends and family. However, infiltrating this network of trust is malicious activity such as sharing hyperlinks to users under the guise of a safe URL, the link may contain malware, or direct the user to a phishing ploy to obtain account details (Himelein-Wachowiak et al., 2021).

There is little doubt that from increased use of social media platforms emerged a more connected global community which, in many ways, has had a significant impact on the awareness of world issues, such as climate change, and allowed individuals to collaborate and build movements to change the world in positive ways. However, on the other side of the coin, it can be dangerous when this information and corresponding movements are built upon a foundation of false information. It could perhaps be argued that this ever-growing global network has been exploited by users wishing to spread propaganda and misinformation. India, for example, which has 200 million monthly WhatsApp users, has experienced several instances of vigilante justice, whereby rumours of false crimes committed by innocent victims have gone viral on WhatsApp, resulting in their murders (de Freitas Melo et al., 2019). Section 4 of this chapter will discuss the impact of the spread of mis- and disinformation in further detail. A report by Kreps et al. (2020) presents that it is perhaps the structure and business models of social media platforms which seek to constantly analyse consumer behaviour that indirectly feeds into the problem (Kreps, 2020). The report argues that ultimately, we live in a world where netizens want instant gratification, which is provided through the personalisation of the content which appears online. This data is then harnessed by social media companies to tailor what users see to keep engagement on the platform high, regardless of the user's immediate interests and disinterests (Kreps, 2020). The report also argues that it is exactly this personalisation which has now been repurposed. No longer just a marketing strategy, political organisations personalise content for users to influence opinions and decision-making. Creating content to increase confusion and political divides can be labour-intensive at best for any human aiming to consistently churn it out, thus bots are employed to complete this task efficiently.

4 The Global Impact of the Spread of Disinformation

It could be argued that, on a micro-level, the effects of coming across mis- and disinformation are often ignored, partly due to the fact that we may not recognise false content when we see it, and therefore may not register it at all. However, it is perhaps interesting to consider the wider impact the spread of mis- and disinformation has had on, not only individuals and immediate networks, but on society and the global community. Interestingly, there exists some debate as to how significant the issue of misinformation actually is in terms of fake news content, with several authors presenting the viewpoint that despite the increased attention on the topic following the 2016 US presidential election and Brexit, fake news is not particularly prevalent online when compared to the magnitude of mainstream news available to the public online, particularly via social media platforms such as Twitter and Facebook (Watts et al., 2021). One such study, 'Measuring the news and its impact on democracy' (2021), estimates that the low impact of disinformation is due to the overall low engagement with online news more generally, finding that in the USA, individuals are five times more likely to consume news via television rather than online, while three quarters of Americans spend less than 30 seconds reading online news sources per day (Watts et al., 2021). The study argues that a broader overview must be taken into account when considering the impact of false information, which should be considered as a wider phenomenon which is not online and includes falsehoods posted online, but also incorrect information broadcast by news channels based on inaccurate information or lies disseminated by politicians for political motives.

Perhaps contrasting this view, it could be argued that it is not necessarily the rate of consumption of disinformation that poses the risk, but rather the mere existence of these falsehoods, combined with the complexity that comes with trying to distinguish between that which is true and that which is untrue. It reflects the instability of trust, which in itself possesses the ability to create fractures in systems and increase confusion and mistrust. It could also be argued that if misinformation is, in fact, overrepresented by recent news and research, this also may create a scenario in which individuals begin to question accurate reliable content, which may also become dangerous. For example, the term 'fake news' was used by former US President Trump and his supporters to describe accurate news stories which reflected badly on him, thus creating confusion as to the meaning of the term (Brown, 2019). In contrast, in a research conducted by Brown (2019), it is argued that misinformation prevents effective decision-making of citizens (Brown, 2019). 'The fake news audience is small and comprises a subset of the Internet's heaviest users, while the real news audience commands a majority of the total internet audience' (Nelson & Taneja, 2018). Effectively, the study presents that misinformation and propaganda are equally destructive. This is given that many dangerous consequences of disseminated falsehoods, such as support for the Iraq war were based on misinformation in the form of misleading statements, not necessarily outright fabrications in order to sway public opinion from one mindset to another (Nelson & Taneja, 2018). Following this theory, the significance of the malicious intent separating misinformation from disinformation may be meaningless, although this is also widely contested.

It is also argued that although statistically the audience for false news stories is relatively limited, second-hand disinformation should also be considered, which occurs as individuals increasingly turn to social media platforms to engage in online news stories (Van Duyn & Collier, 2019). Misinformation carries the ability to tarnish the reputations of individuals, businesses, and alter public decisions as well as decrease the trust the public has in the media more generally (Tandoc et al., 2019; Diehl et al., 2018). According to Arthur C. Clarkes' Third Law, 'any sufficiently advanced technology is indistinguishable from magic' (Clarke, 1973). This goes to say that in a modern, technologically advanced world, it is more complex than ever to look at information and distinguish between truth and falsehoods. It could be argued that with the increase in both mis- and disinformation available online, social media platforms should be subject to the same levels of scepticism and awareness as other sources of information in everyday life, such as news agencies (Diehl et al., 2018). In turn, perceiving the online world in this way may allow government agencies to enforce further security measures to protect individuals online to the same degree as one would be protected in the offline world. Moreover, it is perhaps interesting to consider that the fundamental democratic principles which embodied the rise of the internet, and which allowed netizens to freely construct belief systems on a never-ending spectrum of subjects are now being threatened by the consequences of those very same online freedoms.

5 Mitigating the Impact of Mis- and Disinformation

Watts et al. (2021) argue that to improve the current situation and intercept and prevent the consequences of the spread of misinformation on democracy, a mass collaboration model is required, whereby insights could be reached quicker and with the combined skillsets and insights brought by a multidisciplinary pool of researchers (Watts et al., 2021). Equally, the author presents the idea of the power of a robust system for disseminating the insights and results to third party stakeholders who do not belong to the academic community. In this way, an open research agenda should be shared to ensure that data is readily available for researchers to use. This is one such way to solve the issue of the spread of misinformation on a much wider scale than is currently being explored. It is necessary to collaborate across multiple disciplines because, as it currently stands, research in the field of mis- and disinformation is carried out from a variety of perspectives and under different frameworks, making it difficult to obtain and contrast the research in a meaningful way. The importance of a collaborative effort to successfully tackle the spread of disinformation is shared by the EU Commission's 2018 report (updated in 2020) which sets out actions, including enhanced collaboration with both industry and social media platforms to tackle disinformation.

This includes a call for social media companies to actively: 'Close down fake accounts active on their service, identify automated bots and label them accordingly, and collaborate with the national audio-visual regulators and independent fact-chequers and researchers to detect and flag disinformation campaigns' (de Cock Buning, 2018). It has also been argued that increased education on the subject of misinformation and improved media literacy is required for internet users to reaffirm the importance of awareness of the issue and thus prepare users for not only the dangers of certain forms of misinformation, but also how they may be identified (de Cock Buning, 2018). Awareness for users may include methods in which individuals can mitigate misinformation through their actions when they see a post containing false information, such as not engaging with the content, reporting and blocking the user, and instead sharing official advice to counter the false content (Center for Countering Digital Hate, 2022a). Perhaps another way to increase awareness and to mitigate the issue is to apply further pressure on social media platforms to increase expenditure on security mechanisms to improve how they are able to tackle and detect disinformation online and thus protect users (Pourghomi et al., 2017).

Social media as an ideology is based on the premise of sharing information with multiple users within their network, with the ability to 'go viral', i.e. allow content to be easily and rapidly shared across networks, thus provoking large-scale engagement with the content (Berger & Milkman, 2013). It has frequently been reported that social media companies could do more to mitigate the risks associated with the spread of misinformation. For example, stricter disabling of the sharing mechanism, and increased budgeting and resources for user security. According to an internal memo from Facebook, which was disclosed by the New York Times (The New York Times, 2021), the company is aware that if they were to do nothing, the nature of the algorithms that largely contribute to the make-up of the platform promote the spread of misinformation. Facebook have previously addressed that there are indeed a select group of stakeholders that would require to take on more responsibility in terms of providing solutions (Weedon et al., 2017). In addition, Mark Zuckerberg later announced that Facebook have 40,000 members of staff who work in security of the platform (Facebook, 2022). It could be argued, however, that with 2.9 billion monthly users on the site (Dean, 2022), and therefore one staff member assigned to tackle the content of over 72,500 users, this is insufficient. Social media platforms have made concerted efforts to publish strategies on how to tackle the disinformation on topics such as climate change and COVID-19, where falsehoods are commonly shared online and have the potential to be damaging to public health. One method involves assigning a warning to a post if it is deemed to contain false information. However, according to a study carried out by the Center for Countering Digital Hate (CCDH), although Facebook began tagging posts with information labels and directing users to its Climate Science Information Center, 50.5% of the most popular posts in the sample, posted by far-reaching climate denial news websites, were not tagged (Center for Countering Digital Hate, 2022b).

It is argued that this may lead to the 'implied truth effect', which is described as 'whereby false headlines that fail to get tagged are considered validated and thus are seen as more accurate' (Pennycook et al., 2019). The study conducted by Pennycook et al. considered that the implied truth effect may be mitigated by removing any doubt as to whether a headline had been verified by labelling them as to their verification status. NATO researchers claimed that 'Overall social media companies are experiencing significant challenges in countering the malicious use of their platforms' (Choraś et al., 2021). As such, it could perhaps be argued that tighter restrictions placed on the information that can be shared would be welcome to mitigate sharing of falsehoods on the platform. However, this is complex given that the current system demonstrates a time lag between the content being flagged by a user and the checks that go on beneath the surface (Pourghomi et al., 2017). In addition, it must be acknowledged that the International Grand Committee (IGC) previously concluded that social media companies must not be held solely responsible for acting on the spread of falsehoods on respective platforms (Choraś et al., 2021). It is argued that although Facebook has taken a step in the right direction by implementing the likes of the right-click authenticate process, there are many failings which should be addressed (Pennycook et al., 2019). One way to do this could be to create an advanced ML model across social media platforms which would work to tackle the spread of misinformation on all platforms, instead of specifically one. By focusing efforts of the top companies on one overarching solution, this would be effective to protect users globally.

6 Conclusion

This chapter has presented that there is undoubtedly a complex relationship between falsehoods disseminated through AI methods such as bots, and humans who engage in content and disseminate it within their networks. As such, it appears that mis- and disinformation are interlinked with the online world, which alarmingly may then be spread further by humans offline. With this in mind, it is therefore evident that the issues posed by the spread of both mis- and disinformation are multifaceted and adapting rapidly in correlation with advancements in technology. Thus, this chapter cannot propose to solve these issues outright, but instead merely highlight recommendations to be considered by both academic communities and industry.

This chapter firstly considered the meaning of truth in a post-truth world, exploring various theories pertaining to what constitutes information in its ideal form, before delving into the definitions and variance in the terms 'misinformation' and 'disinformation' in an attempt to expel any discrepancies in their use. Following on from this, the role of AI in the dissemination of falsehoods was explored in Sect. 3, by discussing the methods of fabricating content and how GAN images, deepfakes, and false multimodal content are being identified and studied. The use and types of bots predominantly used online were explored, before contextualising their usage on social media platforms and how humans interacting with content disseminated by bots widens their reach and potential impact. Section 4 then considered the impacts of the spread of mis- and disinformation, including the debate as to the extent of its impact on democracy, and finally, Sect. 5 outlined how the impacts could potentially be reduced through a mass-collaboration framework which would allow research across multiple disciplines to be shared and approached on a wider platform. It is also considered that the burden of responsibility placed on social media companies could perhaps be increased, and that increased media literacy as well as further research into how AI itself could play its part in fighting the spread of falsehoods online could be highly beneficial. Despite the robust body of research which has already and is currently being conducted on the subject, it is crucial that further research is carried out and industry enforces a much stricter plan of action on how to tackle the spread of mis- and disinformation to reduce potential harm to users and thus, wider society.

References

Arendt, H. (1967). *Truth and politics*. The New Yorker.

Bakardjieva, M. (2005). *Internet society: The internet in everyday life*. SAGE Publications.

Benedek, W., & Kettemann, M. C. (2020). *Freedom of expression and the internet: Updated and revised* (2nd ed.). Council of Europe.

Berger, J., & Milkman, K. L. (2013). Emotion and virality: What makes online content go viral? *Insights, 5*, 18–23.

Brown, E. (2019). Propaganda, misinformation, and the epistemic value of democracy. *A Journal of Politics and Society, 30*, 194–218.

Buckland, M. K. (1991). Information as thing. *Journal of the American Society for Information Science, 42*, 351–360.

Bufacchi, V. (2021). Truth, lies and tweets: A consensus theory of post-truth. *Philosophy & Social Criticism, 47*, 347–361.

Burkhardt, J. M. (2017). *Combating fake news in the digital age. Library technology reports* (pp. 5–33). ALA TechSource.

Carsten Stahl, B. (2006). On the difference or equality of information, misinformation, and disinformation: A critical research perspective. *Informing Science Journal, 9*, 83–96.

Center for Countering Digital Hate. (2022a, July 3). *Home: Center for countering digital hate*. Retrieved from Center for Countering Digital Hate Website: https://www.counterhate.com/

Center for Countering Digital Hate. (2022b, February 23). *Facebook failing to flag harmful climate misinformation, new research finds*. Retrieved from Center for Countering Digital Hate Website: https://www.counterhate.com/post/facebook-failing-to-flag-harmful-climate-misinformation-new-research-finds

Chen, Z., Tanash, R. S., Stoll, R., & Subramanian, D. (2017). Hunting malicious bots on Twitter: An unsupervised approach. In *International conference on social informatics* (pp. 501–510). Springer.

Choraś, M., Demestichas, K., Giełczyk, A., Herrero, Á., Ksieniewicz, P., Remoundou, K., ... Woźniak, M. (2021). Advanced ML techniques for fake news (online disinformation) detection: A systematic mapping study. *Applied Soft Computing, 101*, 1–22.

Chu, Z., Gianvecchio, S., Wang, H., & Jajodia, S. (2012). Detecting automation of twitter accounts: Are you a human, bot, or cyborg? *IEEE Transactions on Dependable and Secure Computing, 9*, 811–824.

Ciampaglia, G. L., Mantzarlis, A., Maus, G., & Menczer, F. (2018). Research challenges of digital misinformation: Toward a trustworthy web. *AI Magazine, 39*, 65–74.

Clarke, A. C. (1973). *Profiles of the future*. Harper & Row Publishers. Retrieved from NewScientist: https://www.newscientist.com/definition/clarkes-three-laws/

#:~:text=But%20perhaps%20the%20best%20known%20of%20Clarke's%20three,"Any%20sufficiently%20advanced%20technology%20is%20indistinguishable%20from%20magic"

de Cock Buning, M. (2018). *A multi-dimensional approach to disinformation: Report of the independent high level group on fake news and online disinformation*. Publications Office of the European Union.

de Freitas Melo, P., Coimbra, V., Garimella, K., Vaz de Melo, P. O., & Benevenuto, F. C. (2019). Can WhatsApp counter misinformation by limiting message forwarding? In *International conference on complex networks and their applications* (pp. 372–384). Springer.

Dean, B. (2022, January 5). *Facebook demographic statistics: How many people use Facebook in 2022?* Retrieved from Backlinko Website: https://backlinko.com/facebook-users

Diehl, T., Barnidge, M., & de Zuniga, G. (2018). Multi-platform news use and political participation across age groups: Toward a valid metric of platform diversity and its effects. *Journalism & Mass Communication Quarterly, 96*, 428–451.

El Naqa, I., & Murphy, M. J. (2015). What is ML? In I. El Naqa, M. J. Murphy, & R. Li (Eds.), *ML in radiation oncology* (pp. 3–11). Springer.

European Parliament. (2015, November). *Understanding propaganda and disinformation.* Retrieved from European Parliament Website: https://www.europarl.europa.eu/RegData/etudes/ATAG/2015/571332/EPRS_ATA(2015)571332_EN.pdf

Facebook. (2022, March 7). *Promoting safety and expression*. Retrieved from About Facebook Website: https://about.facebook.com/actions/promoting-safety-and-expression/

Ferrara, E., Varol, O., Davis, C., Menczer, F., & Flammini, A. (2016). The rise of social bots. In *Communications of the ACM* (pp. 96–104). ACM.

Fetzer, J. H. (2004). Disinformation: The use of false information. *Minds and Machines, 14*, 231–240.

Floridi, L. (2004). Outline of a theory of strongly semantic information. *Minds and Machines, 14*, 197–221.

Floridi, L. (2006). The logic of being informed. *Logique et Analyse, 49*, 433–460.

George, S. (2019, June 13). *'Deepfakes' called new election threat, with no easy fix*. Retrieved from AP News: https://apnews.com/article/nancy-pelosi-elections-artificial-intelligence-politics-technology-4b8ec588bf5047a981bb6f7ac4acb5a7

Giorgi, S., Ungar, L., & Schwartz, H. A. (2021). Characterizing social spambots by their human traits. *Findings of the Association for Computational Linguistics: ACL-IJCNLP 2021*, 5148–5158.

Gragnaniello, D., Cozzolino, D., Marra, F., Poggi, G., & Verdoliva, L. (2021). Are GAN generated images easy to detect? A critical analysis of the state-of-the-art. In *2021 IEEE international conference on multimedia and expo (ICME)* (pp. 1–6). IEEE.

Güera, D., & Delp, E. (2018). Deepfake video detection using recurrent neural networks. In *2018 15th IEEE international conference on advanced video and signal based surveillance (AVSS)* (pp. 1–6). IEEE.

Guess, A., & Lyons, B. (2020). Misinformation, disinformation and online propaganda. In N. Persily & J. A. Tucker (Eds.), *Social media and democracy* (pp. 10–33). Cambridge University Press.

Habermas, J. (1984). *The theory of communicative action: Reason and the rationalization of society. Reason and the rationalization of society*. Wiley.

Habermas, J. (2003). *Truth and justification*. Polity Press.

Himelein-Wachowiak, M., Giorgi, S., Devoto, A., Rahman, M., Ungar, L., Schwartz, H. A., ... Curtis, B. (2021). Bots and misinformation spread on social media: Implications for COVID-19. *Journal of Medical Internet Research, 23*(5), e26933.

Horne, B. D., & Adali, S. (2017). This just in: Fake news packs a lot in title, uses simpler, repetitive content in text body, more similar to satire than real news. *arXiv:1703.09398*.

Kreps, S. (2020). *The role of technology in online misinformation*. Brookings.

Kreps, S., McCain, R. M., & Brundage, M. (2020). All the news that's fit to fabricate: AI-generated text as a tool of media misinformation. *Journal of Experimental Political Science, 9*, 104–117.

Lazer, D., Baum, M., Benkler, Y., Berinsky, A., Greenhill, K., & Menczer, F. (2018). The science of fake news: Addressing fake news requires a multidisciplinary effort. *Science, 359*, 1094–1096.

Li, Y., Chang, M.-C., & Lyu, S. (2018). In ictu oculi: Exposing AI created fake videos by detecting eye blinking. In *2018 IEEE international workshop on information forensics and security (WIFS)*. IEEE.

Mingers, J. (2001). Embodying information systems: The contribution of phenomenology. *Information and Organization, 11*, 103–128.

Neelam, M. (2022). Aspects of AI. In J. Karthikeyan, T. Su Hie, & N. Yu Jin (Eds.), *Learning outcomes of classroom research* (pp. 250–256). L'Ordine Nuovo.

Nelson, J. L., & Taneja, H. (2018). The small, disloyal fake news audience: The role of audience availability in fake news consumption. *New Media & Society, 20*, 3720–3737.

Norris, A. (1996). Arendt, Kant, and the politics of common sense. *Polity, 29*, 165–191.

Pedro Baptista, J., & Gradim, A. (2021). "Brave new world" of fake news: How it works. *Journal of the European Institute for Communication and Culture, 28*, 426–442.

Pennycook, G., Bear, A., Collins, E., & Gertler Rand, D. (2019). The implied truth effect: Attaching warnings to a subset of fake news headlines increases perceived accuracy of headlines without warnings. *Management Science, 66*, 4944–4957.

Pourghomi, P., Safieddine, F., Masri, W., & Dordevic, M. (2017). How to stop spread of misinformation on social media: Facebook plans vs. right-click authenticate approach. In *2017 international conference on engineering & MIS (ICEMIS)* (pp. 1–8). IEEE.

Russell, S., & Norvig, P. (1995). *AI: A modern approach*. Prentice-Hall.

Shao, C., Ciampaglia, G. L., Varol, O., Yang, K.-C. F., & Menczer, F. (2018). The spread of low-credibility content by social bots. *Nature Communications, 9*(1), 4787.

Shu, K., Bhattacharjee, A., Alatawi, F., Nazer, T. H., Ding, K., Karami, M., & Liu, H. (2020). Combating disinformation in a social media age. *WIREs Data Mining and Knowledge Discovery, 10*, 1–23.

Simon, H. A. (1983). Why should machines learn? In R. S. Michalski, J. G. Carbonell, & T. M. Mitchell (Eds.), *ML: An AI approach* (pp. 25–37). Elsevier Inc.

Singh, V. K., Ghosh, I., & Sonagara, D. (2020). Detecting fake news stories via multimodal analysis. *Journal of the Association for Information Science and Technology, 72*, 3–17.

Tandoc, E. C., Jr. (2019). The facts of fake news: A research review. *Sociology Compass, 13*(2), 1–10.

Tandoc, E. C., Lim, D., & Ling, R. (2019). Diffusion of disinformation: How social media users respond to fake news and why. *Journalism, 21*, 381–398.

The New York Times. (2021, October 25). *Facebook wrestles with the features it used to define social networking*. Retrieved from The New York Times Website: https://www.nytimes.com/2021/10/25/technology/facebook-like-share-buttons.html

Turing, A. M. (1950). Computing machinery and intelligence. *Mind, 59*, 433–460.

Van Duyn, E., & Collier, J. (2019). Priming and fake news: The effects of elite discourse on evaluations of news media. *Mass Communication and Society, 22*, 29–48.

Vasu, N., Ang, B., Teo, T.-A., Jayakumar, S., Faizal, M., & Ahuja, J. (2018). *Fake news: National security in the post-trust era*. S. Rajaratnam School of International Studies.

Watts, D. J., Rothschild, D. M., & Mobius, M. (2021). Measuring the news and its impact on democracy. In *Advancing the science and practice of science communication: Misinformation about science in the public sphere* (pp. 1–6). PNAS.

Webster, F. (2014). *Theories of the information society*. Routledge.

Weedon, J., Nuland, W., & Stamos, A. (2017, April 27). *Information operations and Facebook*. Retrieved from About Facebook Website: https://about.fb.com/br/wp-content/uploads/sites/3/2017/09/facebook-and-information-operations-v1.pdf

Yuezun, L., Chang, M.-C., & Lyu, S. (2018). Exposing AI generated fake face videos by detecting eye blinking. In *IEEE international workshop on information forensics and security (WIFS)* (pp. 1–7). IEEE.

Zhang, X., & Ghorbani, A. A. (2020). An overview of online fake news: Characterization, detection, and discussion. *Information Processing & Management, 57*, 2–26.
Zhang, X., Karaman, S., & Shih-Fu, C. (2019). Detecting and simulating artifacts in GAN fake images. In *2019 IEEE international workshop on information forensics and security (WIFS)* (pp. 1–6). IEEE.

How States' Recourse to Artificial Intelligence for National Security Purposes Threatens Our Most Fundamental Rights

Océane Dieu and Reza Montasari

Abstract Many states deploy artificial intelligence (AI) and associated technology in their efforts to safeguard their national security. When these states justify their recourse to AI for national security purposes by arguing that 'technology and machines are neutral', they disregard one essential element: technology is far from neutral. Inherent biases and errors in AI deployed in national security uses seriously threaten people's fundamental rights. Citizens subjected to intrusive AI-enabled technology see, amongst others, their right to privacy, right to a fair trial, right to freedom of opinion and even their most fundamental right to life endangered. This work seeks to investigate and raise awareness regarding several human rights threatened by the state's recourse to AI for national security purposes.

Keywords Artificial intelligence · National security · Surveillance · European Convention on Human Rights · Right to privacy · Freedom of expression · Right to life · Fair trial · Unmanned aerial vehicles · Lethal autonomous weapons · Foreign state disinformation

1 Introduction

That authoritarian regimes, such as China, largely rely on surveillance is unlikely to surprise the reader. However, China is not the only country that extensively deploys surveillance technology. Western democracies, including France, the United Kingdom and the United States are increasingly developing and implementing surveillance technology to keep their citizens and country 'safe' (Feldstein, 2019).

O. Dieu (✉)
Hillary Rodham Clinton School of Law, Swansea University, Swansea, UK

R. Montasari
School of Social Sciences, Department of Criminology, Sociology and Social Policy, Swansea University, Swansea, UK
e-mail: Reza.Montasari@Swansea.ac.uk
URL: http://www.swansea.ac.uk

R. Montasari (ed.), *Artificial Intelligence and National Security*,
https://doi.org/10.1007/978-3-031-06709-9_2

Relying on safety and security reasons is problematic when states try to legitimise their widespread recourse to surveillance under the pretext of national security. This recourse becomes even more worrisome when it is justified by a 'technology is neutral' discourse (European Union Agency for Fundamental Rights, 2020, p. 28). Technology is human-made, and humans are all but neutral. Consequently, it would be unrealistic to think humans can make neutral technology.

While state surveillance is not unlawful and its use has legitimate purposes, the impact of these far-reaching surveillance techniques on citizens' fundamental rights is worrisome. This work seeks to give an overview of several human rights endangered by states' recourse to artificial intelligence (AI) for national security purposes. The decision was made to limit this analysis to the threats such recourse can constitute to, on the one hand, human rights at the level of the Council of Europe, and, on the other hand, international humanitarian law. Consequently, the human rights' legal framework at the international level, the European Union level or the national level will not be analysed. Moreover, whilst data protection law is very relevant to this analysis, it will not be covered by this work. Furthermore, this paper does not aim to provide an exhaustive overview of all the domains in which states deploy AI for national security purposes, but the choice was made to limit the discussion to states' surveillance practices, the deployment of drones and lethal autonomous weapons, foreign state disinformation and the online fight against illegal content. To exemplify this rather theoretical analysis, some of the United Kingdom's (UK) practices will be integrated.

The remainder of this chapter is structured as follows. Section 2 discusses preliminary notions and remarks aiming to set the study in context. Section 3 presents an overview of the legal framework of some of the human rights at stake. Section 4 critically examines the concrete legal, human rights and ethical issues that arise with the states' deployment of AI for national security purposes. More specifically, Sect. 4.1 investigates the threats of AI in national security uses in general while Sect. 4.2 examines states' recourse to AI in surveillance techniques. Section 4.3 explores the development of AI in drones and lethal autonomous weapon systems. Section 4.4 analyses the implication of AI in foreign state-led disinformation campaigns whereas Sect. 4.5 discusses the adoption of AI in the fight against illegal content through the detection of online threats. Finally, this chapter is concluded in Sect. 5.

2 Background

2.1 National Security Being a Vague and Ambiguous Term

The notion of 'national security' lacks a clear, exhaustive and unambiguous definition. At the level of the United Kingdom, the Information Commissioner's Office understands 'national security' as 'the security and well-being of the UK as a whole, its population, and its institutions and system of government', covering amongst others the protection against terrorist threats (Information Commissioner's Office, n.d.). As such, the British security service MI5's role is to protect the national security of the Kingdom, and in particular against threats of 'terrorism, espionage and sabotage, the activities of agents of foreign powers, and from actions intended to overthrow or undermine parliamentary democracy by political, industrial or violent means' (Security Service Act 1989, s.1(2)). At the level of the Council of Europe, the Convention for the Protection of Human Rights and Fundamental Freedoms 1950 (the European Convention on Human Rights, the Convention or the ECHR) protects the most fundamental rights of the citizens of the Convention's member states. When these citizens consider that their human rights have been violated by a member state, they can seek judicial redress with the European Court of Human Rights (ECHR, art. 19). Whilst a definition of the notion of national security is absent at the level of the European Court of Human Rights, which considers this to fall within the states' margin of appreciation, the Court has delimitated the notion through its case-law. Overall, the notion includes 'the protection of state security and constitutional democracy from espionage, terrorism, support for terrorism, separatism and incitement to breach military discipline' (European Court of Human Rights, 2013, p. 5). However, it is questionable whether this definition can justify the states' extensive recourse to AI in national security uses.

The protection of national security is one of the reasons why states can restrict their citizens' qualified human rights that are legally enshrined in the European Convention on Human Rights. The Convention distinguishes *qualified* human rights, that are to be understood as 'rights which may be interfered with in order to protect the rights of another or the wider public interest' (Council of Europe, n.d.), from *absolute* human rights that are rights from which states can under no circumstance derogate (Council of Europe, n.d.). The Convention contains several qualified rights relevant to the current analysis of the recourse by states to AI for national security purposes. As such, the right to respect for private and family life, the right to freedom of expression and the right to freedom of assembly and association are rights that can be restricted, within certain limits, for national security purposes. These qualified rights can be restricted when such a restriction is provided by a clear, accessible and foreseeable law, is necessary in a democratic society and pursues a legitimate aim, such as, amongst others, the protection of national security (ECHR, art. 8(2); art. 10(2); art. 11(2)) (hereinafter this three-levelled test will be referred to as the 'legality-legitimacy-necessity test'). If the restriction of the person's rights stands this test, the interference with their right is legitimate and, therefore, allowed. Absolute human rights, such as the right to life or the prohibition of torture, respectively, protected by articles 2 and 3 of the Convention, can under no circumstance be restricted.

National security and human rights are sometimes considered to be mutually exclusive (Schwartz, 2014). Taking as an example the global war against terrorism, the restriction of citizens' most fundamental human rights that ensues this fight, has often been justified due to national security reasons (da Silva, 2017; Koehler-Schindler, 2021; United Nations Office of the High Commissioner for Human Rights, 2008). However, the fight against terrorism is often used as an excuse to impede those citizens' human rights beyond that what is necessary to fight real terrorist threats. Consequently, citizens' rights are unnecessarily restricted in the name of this fight. However, the rationale behind these rights is the protection of citizens against states' excessive restriction of their fundamental rights. As Lord Phillips of Worth Matravers, former President of the UK Supreme Court, stated: 'The so called "war against terrorism" is not so much a military as an ideological battle. Respect for human rights is a key weapon in that ideological battle' (Lord Phillips of Worth Matravers, 2010). It is, therefore, of utmost importance that states adopt a balanced approach in their recourse to AI for national security purposes when that recourse leads to a restriction of their citizens' fundamental rights.

2.2 Understanding Artificial Intelligence

2.2.1 The Notion of Artificial Intelligence

AI, algorithms and machine learning (ML) are terms that are often used intertwiningly. Hence, a clear delimitation of these terms is required to fully understand states' use of these technologies for national security purposes. In the absence of a universally agreed-upon definition of AI (Babuta et al., 2020), the definitions provided by two of AI's founding fathers, John McCarty and Marvin Minsky, will serve as the basis for the following analysis. John McCarty defined 'AI' as the 'science and engineering of making intelligent machines' (McCarthy, n.d.) and Marvin Minsky as 'the science of making machines do things that would require intelligence if done by [humans]'[1] (World Commission on the Ethics of Scientific Knowledge and Technology, 2017, p. 17). Traditionally, two types of AI can be distinguished: narrow AI and general AI. *Narrow* AI can be understood as the 'single-task application of AI for uses' (Access Now, 2018, p. 8), such as AI-operated autonomous vehicles or online translation tools (Access Now, 2018; United Nations Special Rapporteur on the promotion and protection of the right to freedom of opinion and expression, 2018). Whilst *narrow* AI is capable of performing one computational task (Szocik & Jurkowska-Gomułka, 2021), *general* AI attains the same level of intelligence as the human brain, if not superior, and can perform several cognitive tasks (Access Now, 2018; United Nations Special Rapporteur on the promotion and protection of the right to freedom of opinion and expression, 2018). As the current technology stands, general AI is still something of a science fiction movie since scientists have not yet been able to create such technology (Babuta et al., 2020).

[1] The original definition reads 'the science of making machines do things that would require intelligence if done by "men"'. For the sake of gender-inclusiveness, 'men' has been changed to 'humans'. The same remark can be made regarding rulings of the European Court of Human Rights, where the Court's exclusive reference to 'man', 'his' and other male inclinations have been replaced by the more gender-inclusive forms 'humans', 'persons' and 'their'.

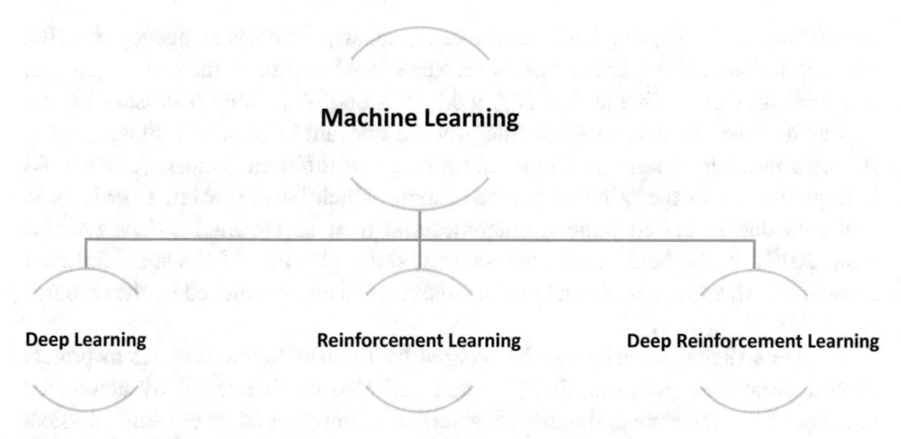

Fig. 1 Different sub-categories of ML based on Haney's (2020) study

Algorithms can be defined as 'a set of mathematical instructions or rules that, especially if given to a computer, will help to calculate an answer to a problem' (Cambridge Dictionary, n.d.). Whilst AI functions through algorithms, such as neural networks used in deep learning, not all algorithms implicate AI (Access Now, 2018). ML is an AI mechanism that allows an algorithm to learn from its previous decisions to improve its performance (Jorgensen, 2019; Surden, 2014). ML is a sub-discipline of narrow AI models that is relevant to the national security uses of AI. More specifically, there exist three sub-categories of ML that are used in this field (Haney, 2020). Figure 1 provides a graphical representation of these sub-categories.

Deep learning can be understood as 'a process by which neural networks learn from large amounts of data' (Haney, 2020, p. 64) and functions in a similar way as the human brain operates. Put simply, the neural networks are fed with large amounts of data ('the input layer'), which the model processes to attain a certain outcome ('the output layer') (Haney, 2020). *Reinforcement learning* can be understood as 'a type of machine learning concerned with learning how an agent should behave in an environment to maximize a reward' (Haney, 2020, p. 68). *Deep reinforcement learning* is a combination of the two previous categories of machine learning (Haney, 2020). Given that deep learning is a form of machine learning, which in turn is an application of AI, the umbrella term 'AI' will equally be referred to when discussing these more specific techniques.

2.2.2 Threats and Opportunities of Artificial Intelligence

Whilst this work particularly focuses on the legal, human rights and ethical threats AI can constitute, it is worth highlighting that it also presents many opportunities. As such, AI can facilitate humans' daily life, by, for example, helping persons with visual disabilities or optimising agriculture (Access Now, 2018). Opportunities also exist in the national security scene. Given the data overload and complexity of online data (Oswald, 2020), the UK's intelligence, security and cyber agency, GCHQ, has emphasised the importance of its recourse to AI to protect the country against any malicious cyber-threats (GCHQ, n.d.). It is unquestionably necessary for the agency to have recourse to AI to filter out the relevant information. States deploy AI for a number of reasons and within a range of different sectors. As such, AI is implemented in the criminal justice system, which has, however, heavily been criticised due to biased judicial outcomes and their severe implications (Access Now, 2018). In the field of national security, states also use AI. As such, different aspects of a state's security can both be strengthened and threatened by the recourse to AI.

A state's digital security can be secured by the development of AI to protect its infrastructures (Johnson, 2019), but it can also be threatened by automated hacking of foreign state and non-state actors. A country's and its citizens' physical security can also be endangered, by, for example, the deployment of AI-operated lethal autonomous weapons that kill humans or unmanned aerial vehicles (otherwise known as drones) that can cause physical injury when falling onto people. Moreover, a country's political security can also be endangered by the use of AI, by, for example, foreign disinformation techniques. It can, however, also be secured by deploying surveillance techniques (Johnson, 2019; Yu & Carroll, 2021). Whilst the recourse to AI can both endanger and strengthen a state's national security, such recourse can also threaten its citizens' fundamental rights. As such, citizens' right to privacy is endangered when states use invasive drones or surveillance techniques. These innovative technologies can become the object of hacking, allowing the hackers to obtain access to that data or use the data fraudulently. The use of drones and surveillance techniques are only a few examples of how the recourse to technology and artificial intelligence by states can become problematic. This recourse becomes even more controversial when it is done under the cover of protecting national security. This work acknowledges the growing importance of the development of AI in national security uses but also recognises the threats it can constitute to human rights. Consequently, the purpose is not to outright set aside all uses of AI in a national security context but to offer a more nuanced approach where national security is balanced with fundamental rights.

3 Legal Framework of Endangered Human Rights

Several fundamental human rights are endangered by states' use of AI in the context of national security. In the following parts, the legal framework of the European Convention on Human Rights regarding the relevant human rights will be set out.[2] Moreover, the choice was made to address only some human rights that are especially relevant to the current analysis.

3.1 *Right to Life*

The first relevant right is the absolute right to life which is enshrined in article 2 of the European Convention on Human Rights. This right is considered to be one of the most basic values of a democratic society (Giuliani and Gaggio v. Italy, 2011, para. 174). Consequently, this right does not permit any derogation under article 15 of the Convention.[3] The right to life contains a positive obligation for states to protect by law and operational measures the life of citizens in their jurisdiction and a prohibition to intentionally deprive their citizens of life, safe for a few exceptions (European Court of Human Rights, 2021a). The state's positive obligation to protect life does not, however, extend to 'guaranteeing to every individual an absolute level of security in any activity in which the right to life may be at stake' (Gülsen Gökdemir v. Turkey, 2015, para. 17; Molie v. Romania, 2009, para. 44).

3.2 *Right to a Fair Trial*

The second important right is enshrined in article 6 of the Convention. This article legally protects one's right to a fair trial. The right to a fair trial covers a range of minimal sub-rights such as the right to be informed promptly of the nature and cause of the accusation, the principle of equality of arms and the presumption of innocence. The right to be informed promptly of the nature and cause of the accusation (ECHR, art. 6(3)(a)) is deeply linked to the general right to a fair hearing of article 6(1) of the ECHR given that the notification of the charges against a person 'is an essential prerequisite for ensuring that the proceedings are fair' (Pélissier and Sassi v. France, 1999, para. 52; Varela Geis v. Spain, 2013, para. 42). The defendant

[2] As stated previously, due to the limited scope of this work, the human rights threats will only be analysed on the basis of the legal framework of the European Convention on Human Rights, with the exclusion of other international instruments, such as the International Covenant on Civil and Political Rights of 1966, European Union instruments, such as the EU Charter of Fundamental Rights of 2012, or national instruments, such as the UK's Human Rights Act 1998. Moreover, other legal frameworks, such as the data protection framework will not be addressed.

[3] Article 15 of the Convention allows states to derogate from certain qualified rights of the Convention in times of war or public emergency when such restrictions are required by the exigencies of the emergency situation. The current analysis of national security uses of AI however does not extend to the analysis of states' state of emergency or times of war, but is restricted to the state's daily use of AI for national security purposes. Hence, the possibility to derogate from those rights will not be discussed.

must receive sufficient information regarding the accusation to fully understand its extent (European Court of Human Rights, 2021b). On the principle of equality of arms, the European Court of Human Rights has stated in several judgments 'that each party must be afforded a reasonable opportunity to present [their] case under conditions that do not place [them] at a disadvantage vis-à-vis [their] opponent' (Foucher v. France, 1997, para. 34; Öcalan v. Turkey, 2005, para. 140). The Court moreover ruled that the equality of arms principle is also violated when 'the accused has limited access to [their] case file or other documents on public-interest grounds' (Matyjek v. Poland, 2007, para. 65; Moiseyev v. Russia, 2009, para. 217). The principle of equality of arms can, however, legitimately be restricted in light of interests of national security (Kennedy v. United Kingdom, 2010, para. 187). Last, the presumption of innocence covers several principles, including the prohibition for a court to start proceedings with 'the preconceived idea that the accused has committed the offence charged' (Barberà, Messegué and Jabardo v. Spain, 1988, 77). Moreover, this presumption puts the burden of proof on the prosecution and requires that 'any doubt should benefit the accused' (Barberà, Messegué and Jabardo v. Spain, 1988, 77).

3.3 Right to Privacy and Data Protection

A third right endangered by national security uses of AI is the right to private and family life. This right is legally enshrined in article 8 of the ECHR, which more specifically recognises a qualified right to respect for one's private and family life, one's home and correspondence. Whilst not explicitly mentioned in the Convention, the European Court of Human Rights has recognised that the right to data protection is of such 'fundamental importance to a person's enjoyment of the respect to private and family life' that this right to data protection falls under the scope of the right to privacy as legally safeguarded by article 8 of the Convention (Satakunnan Markkinapörssi Oy and Satamedia Oy v. Finland, 2017, para. 137; Z v. Finland, 1997, para. 95). This right can be restricted if the restriction passes the legality-legitimacy-necessity test. The right to privacy is strongly linked to the freedom of opinion and expression, given that the former 'often acts as a gateway to the enjoyment' of the latter (United Nations Special Rapporteur on the promotion and protection of the right to freedom of opinion and expression, 2018, p. 13).

3.4 Right to Freedom of Expression and Opinion

The fourth and fifth fundamental rights that are endangered by a states' recourse to AI for national security reasons are the qualified right to freedom of expression and the absolute right to hold opinions. Both rights are legally protected under article 10 of the European Convention on Human Rights. The absolute right to hold opinions falls under the right to freedom of expression as the former is a prerequisite of the enjoyment of the latter. Given the absolute character of this right, states can under no circumstance interfere with it, such as through indoctrination (Bychawska-Siniarska, 2017). The qualified right to freedom of expression prohibits the state from interfering with the citizens' right to freedom of expression.[4] The right to freedom of expression is considered to be the backbone of democracies. Without this fundamental right, citizens cannot fully enjoy and participate in a democratic regime. Given this importance, this right has been recognised to cover a broad range of acts that are not limited to positive speech. As such, the European Court of Human Rights recognised in 1976 that the right to freedom of expression is also applicable to ideas 'that offend, shock or disturb the State or any sector of the population. Such are the demands of that pluralism, tolerance and broadmindedness without which there is no 'democratic society" (Handyside v. UK, 1976; Dieu et al., 2021). Contrary to the absolute right to hold opinions, the qualified right to freedom of expression can be limited. Recognising the responsibilities that accompany this qualified right, article 10(2) of the ECHR allows for restrictions that stand the legality-legitimacy-necessity test. Importantly, the Court has held in several judgments that these restrictions cannot be used to silence political opponents (Brasilier v. France, 2006; Sürek v. Turkey, 1999). Political opponents are sometimes silenced on charges of terrorism (Dieu et al., 2021). Consequently, and given the ambiguous delimitation of the notion of 'national security', a delicate approach is required when limiting a person's right to freedom of expression based on national security allegations.

3.5 *Right to Freedom of Assembly and Association*

The sixth relevant right is the right to freedom of assembly and association that is enshrined in article 11 of the European Convention on Human Rights. This qualified right is limited to peaceful assembly, excluding as such any right to gather by means of violence. Moreover, the article specifies that the armed forces, the police or the administration of a state can impose legal restrictions to this right, without constituting an infringement of the right to freedom of assembly and association (ECHR, art. 11(2)). The right to freedom of assembly and association is linked to the right to freedom of expression since the exercise of one's freedom of association can be seen as the collective exercise of one's freedom of expression. The freedom of assembly provides a forum in which one can exercise their right to freedom of expression (European Court of Human Rights, 2021c).

[4] Besides this prohibition, the right to freedom of expression also contains a positive counterpart that obliges states to ensure that whenever a conflict arises between a citizen and a private company, the citizen will still be able to enjoy this right. See: Jorgensen and Pedersen (2017).

3.6 Right to Equality: The Prohibition of Discrimination

Particularly relevant to the states' use of AI for national security purposes is the prohibition of discrimination, otherwise known as the right to equality. Discrimination can be understood as a situation 'where one person is treated less favourably than another is, has been or would be, treated in a comparable situation, based on a perceived or real personal characteristic' (European Union Agency for Fundamental Rights, 2020, p. 68). The right not to be discriminated against in the enjoyment of the ECHR's rights finds its framework in article 14 of the Convention and prohibits discrimination on grounds of sex, race, colour, language, religion, political or other opinion, national or social origin, association with a national minority, property, birth or other status. Contrary to other rights such as the right to freedom of expression, this article is not an independent right to which the Court can find a violation, but it complements the other Convention's rights. More specifically, a violation of the non-discrimination principle requires a discrimination in the enjoyment of one of the other Convention's rights, such as the right to freedom of expression or the right to privacy. To prove such discrimination, it is, however, not required that the Court finds a violation of the other substantive provision (European Court of Human Rights, 2021d).

3.7 Right to Free Elections

The last relevant right is the right to free elections and was not initially provided for in the Convention, but was added by the first Protocol to the Convention for the Protection of Human Rights and Fundamental Freedoms 1952 (Protocol No. 1). Article 3 of this Protocol provides that the member states will 'hold free elections at reasonable intervals by secret ballot, under conditions which will ensure the free expression of the opinion of the people in the choice of the legislature'. This right to free elections is inherently linked to the right to freedom of expression. As such, the European Court of Human Rights has found that freedom of expression is a 'necessary condition to ensure the free expression of the opinion of the people in the choice of the legislature. For this reason, it is particularly important in the period preceding an election that opinions and information of all kinds are permitted to circulate freely' (Bowman v. the United Kingdom, 1998, para. 42; Mathieu-Mohin and Clerfayt v. Belgium, 1987, para. 54).

4 National Security Uses of Artificial Intelligence

As GCHQ stated, the recourse by states to AI for national security uses in general comes with high concerns of human rights violations and 'unique ethical challenges'

(GCHQ, n.d.). With the importance of the centricity and prevalence of human rights in the context of the recourse to AI for national security purposes in mind, several national security uses of AI and their implication on human rights will be examined. More specifically, issues of biases, errors, false positives and false negatives and problems of transparency and accountability, when states use artificial intelligence, will be discussed. Afterwards, the human rights threat in the states' deployment of AI in surveillance techniques under the cover of 'national security purposes', in the field of predictive policing and combined with facial recognition technology, will be analysed. Furthermore, the recourse to AI-enabled drones and lethal autonomous weapons systems will be addressed. Last, the involvement of AI in foreign state disinformation campaigns and in the online fight against illegal content will be assessed from a human rights point of view.

4.1 States' Recourse to Artificial Intelligence in General

4.1.1 Problems of Biases, Errors, False Positives and False Negatives

A first risk to the state's recourse to AI is the inherent biases algorithms can contain that will eventually lead to errors in outcomes (McKendrick, 2019; Secretary-General of the United Nations, 2018). A bias can be understood as 'an inclination or prejudice for or against a person or group, especially in a way that is considered to be unfair' (Access Now, 2018, p. 11). Biased AI can occur at two stages: at the level of the creation of the system and the input level (Access Now, 2018). AI systems are created by humans who can be biased, for example, because of personal convictions (Amnesty International & Access Now, 2018; Leslie, 2019; Yu & Carroll, 2021). When a data scientist who creates an AI model considers, for example, that all terrorist content is always published by Muslim persons, this 'human-induced' bias (Yu & Carroll, 2021) might be transcribed in the model the person is designing. Consequently, this would create a system that would disproportionately qualify content published by Muslims as terrorist content. The algorithm would then operate on a discriminatory basis, as prohibited under the right to equality (Dieu et al., 2021). At the input level, an AI model can be fed with data that is already biased (Yu & Carroll, 2021). As such, using such biased historical data or data that is not representative of a population or incorrect, outdated or poor quality data, can result in a 'data-driven' biased outcome (Access Now, 2018; Babuta et al., 2020; Leslie, 2019; Yu & Carroll, 2021). Once the technique of machine learning is applied to these models that contain biases, these biases will, due to the feedback loops that reinforce these biases (Access Now, 2018; Committee of Ministers of the Council of Europe, 2020; Leslie, 2019; Osoba & Welser IV, 2017; Yu & Carroll, 2021), only be strengthened and emphasised, creating as such a 'machine self-learning bias' (Yu & Carroll, 2021).

Babuta, Oswald and Janjeva interestingly and rightly point out that biased AI systems 'are likely to be no more [biased] than existing human decision-making processes' (Babuta et al., 2020, p. 29). However, when decision-making processes are entirely reliant on artificial intelligence systems, the errors in these AI tools can have far-reaching consequences for individuals (Babuta et al., 2020). As such, biases and errors can lead to false positive or false negative results, where persons will, respectively, be wrongly qualified as falling within a certain category or erroneously not be qualified as such. When taking facial recognition technology as an example, a false positive will occur when a person's face has wrongly been matched with another person's face contained in a database. A false negative will erroneously not recognise a person's face as matching with the same person's face in a database. Moreover, it is questionable whether all the programmers behind the AI systems or the persons inputting data into an AI programme are aware of their own internal biases or the biases that exist in the data. Consequently, questions of transparency and accountability become highly important in the context of a state's recourse to AI.

4.1.2 Problems of Transparency and Accountability

The issue of transparency is of paramount importance when analysing the national security uses of AI (Rodrigues, 2020). 'Transparency' can be defined as the 'availability of information about an actor allowing other actors to monitor the workings and performance of this actor' (Bovens & Schillemans, 2016, p. 511). Biases in AI and the absence of transparency regarding AI systems raise important human rights concerns on the right to a fair trial (Committee of Experts on Internet Intermediaries, 2018; European Union Agency for Fundamental Rights, 2020). When states deploy deep learning techniques, this absence of transparency becomes apparent. Due to their inherent 'black-box' nature (Babuta et al., 2020), these deep learning techniques obscure the processes and techniques that lead to a certain outcome. The developers of such techniques are often not the ones feeding the programmes with data or prosecuting individuals based on such AI programmes. GCHQ, therefore, emphasises the importance of 'explainable AI', which will allow non-technically skilled users to understand the AI systems (GCHQ, n.d.). This should not be limited to GCHQ's use of AI in decision-making processes but should extend to all governmental sectors. However, when deep learning techniques are used in the process of someone's conviction, it is questionable whether the right to a fair trial can be complied with when the prosecuting party and a non-technically skilled defendant don't fully understand the extent to which these techniques have arrived at a certain outcome that was crucial in the person's conviction.

Moreover, whilst the disclosure of certain evidence to a defendant can be refused on grounds of national security protection (European Court of Human Rights, 2021b), it is also questionable how and to what extent the state would grant access to AI-based decisions that lie at the origin of the prosecution, given the absence of transparency regarding these AI-systems. Consequently, GCHQ's emphasis on 'explainable AI' is a laudable effort to increase transparency that should become a general standard for all governmental decisions. An increase in transparency is often countered by reasons of innovation. The private sector often uses the argument that full transparency of AI prevents industrial innovation and investment (Access Now, 2018; European Union Agency for Fundamental Rights, 2020). However, and agreeing with Access Now, full transparency of AI used for national security purposes would enable the state to comply with its obligations under human rights law. Moreover, it would allow researchers and IT specialists to highlight biases in the AI systems that the creator of the system has overlooked or was unaware of. Besides this possibility for evaluation, feedback and critique, increased transparency would allow formal sanctioning where needed (Moses & Janet, 2018). Consequently, increasing transparency would only elevate the population's trust and the state's accountability for its use of the AI systems in its national security operations (Access Now, 2018). Transparency and accountability are, therefore, intrinsically linked. The Committee of Experts on Internet Intermediaries defines 'accountability' as 'the principle that a person who is legally or politically responsible for harm has to provide some form of justification or compensation' (Committee of Experts on Internet Intermediaries, 2018, p. 39).

Transparency and accountability of a state's recourse to AI in national security uses becomes even more important in areas such as law enforcement, given the drastic consequences law enforcement decisions can have on citizens. Furthermore, as Oswald points out, the 'cumulative risk of AI systems interacting with each other' (Oswald, 2020) requires even higher standards of transparency and accountability, especially when law enforcement agencies have access to interoperable AI-enabled databases. However, it is still very uncertain whether this accountability for errors in AI should lie on the developer, which would go against innovation, or on the state (Heilemann, 2021; Rodrigues, 2020; Szocik & Jurkowska-Gomułka, 2021). Given the problems of biases, errors, false positives and false negatives and the requirements of transparency and accountability, the importance of keeping a 'human-in-the-loop' (Enarsson et al., 2022), and not solely relying on AI systems, in the recourse to AI for national security purposes becomes apparent. However, as GCHQ points out, it is not always easy to keep a human-in-the-loop in AI-enable decision-making given the speed with which this decision-making has to happen (GCHQ, n.d.).

4.2 *Surveillance Practices*

Turning to the more specific uses of AI for national security purposes, surveillance techniques have widely been developed over the years. Surveillance by the state on their citizens has increased significantly due, amongst others, to the development of the internet of things or the recourse to big data (Loideain, 2019). Big data can be understood as 'datasets that are generally extremely large in volume, are collected near or in real time, link different sources or levels of information together, and which contain diverse variables that are detailed and tend to be exhaustive in scope' (Hardyns & Rummens, 2018, pp. 201–202). Not only the scale of surveillance and the amount of data available has increased but surveillance has also become more intrusive (Access Now, 2018). Feldstein's research shows that at least 75 out of 176 countries analysed actively used artificial intelligence in their surveillance practices. Whilst it should not come as a surprise that authoritarian regimes, such as China, extensively use AI in their surveillance mechanisms, many liberal Western countries, such as the United Kingdom, the United States or France, utilise them for, amongst others, the development of smart cities, smart policing and facial recognition purposes (Feldstein, 2019).

4.2.1 Surveillance Practices Limited Only to Legitimate National Security Purposes

State surveillance is legitimate when it is employed for the protection of national security, such as for the prevention of terrorist attacks, and can take different forms, including through the deployment of CCTV, internet monitoring, bulk interception and collection of communications or telephone tapping (European Court of Human Rights, 2022)).[5] As such, the deployment of AI in surveillance can increase cost efficiency and cover a broader territory than traditional surveillance operated by humans (Feldstein, 2019). However, two issues regarding state surveillance rise. First, surveillance technology is not limited to legitimate national security reasons but also extends to public safety purposes. Whilst states' use of surveillance for public safety reasons lies outside of this work's scope, it is highly worrisome that surveillance mechanisms for public safety purposes are deployed under the cover of national security reasons. This unchallenged broadening of the purpose and scope of surveillance for initial purposes of the protection of national security to general public safety, also known as 'function creep', is very problematic, especially when the public is unaware of it (Koops, 2021). As such, states have been deploying large scale public surveillance policies to create 'smart cities' or 'safe cities', where interconnected devices allow states to optimise their city management and service delivery (Feldstein, 2019).

Second, the generalised surveillance in itself is problematic for several reasons, regardless of whether the surveillance is carried out for national security or extended public safety reasons. Bearing in mind the inherent errors and biases AI-enabled technology carries, citizens' fundamental rights are endangered by false positive and false negative results. On the one hand, falsely identifying a person as a wanted person (false positive) can lead to accusing a person of a crime they have not committed. On the other hand, not recognising a wanted person (false negative), causes public safety issues. Besides the inherent interference with citizens' fundamental right to privacy, as protected under article 8 of the ECHR, the constant surveillance of unaware citizens puts them in a weak position (Committee of Experts

[5] The current analysis of states' practices of surveillance will only address the recourse to CCTV, with the exclusion of other forms of surveillance practices.

on Internet Intermediaries, 2018; European Union Agency for Fundamental Rights, 2020). The absence of awareness of being monitored takes away their possibility to question and challenge this constant surveillance. Moreover, constant surveillance can have a chilling effect on citizens' behaviour. When citizens become aware of being monitored, they might adapt their behaviour and movements to this constant surveillance. As such, citizens might want to avoid being detected by surveillance by not exercising their fundamental right to association and assembly anymore. The chilling effect of surveillance measures will then endanger other fundamental rights of the surveyed citizens (Access Now, 2018; European Union Agency for Fundamental Rights, 2020).

Consequently, AI-enabled surveillance mechanisms should be limited only to legitimate national security uses, whilst being particularly aware of the potential human rights violations surveillance practices can engender. As the Special Rapporteur on the promotion and protection of the right to freedom of opinion and expression David Kaye stated, the recourse to surveillance in the interest of national security, restricting consequently the human rights of the surveyed, should be 'limited in application to situations in which the interest of the whole nation is at stake, which would thereby exclude restrictions in the sole interest of a Government, regime or power group' (United Nations Special Rapporteur on the promotion and protection of freedom of opinion and expression, 2019, p. 8).

4.2.2 Surveillance Practices Deployed in Predictive Policing

Data extracted from surveillance mechanisms are used in predictive policing. Predictive policing can be understood as 'the use of historical data to create a spatiotemporal forecast of areas of criminality or crime hot spots that will be the basis for police resource allocation decisions with the expectation that having officers at the proposed place and time will deter or detect criminal activity' (Ratcliffe, 2014, p. 4). Accordingly, by feeding data regarding previous crimes into an AI model, the model is able to predict the location and probability of and the potential individuals likely to commit future crimes (Blount, 2021; European Union Agency for Fundamental Rights, 2020; Feldstein, 2019). Based on data regarding previous crimes collected through massive surveillance, predictive policing will 'predict' future crimes, criminals and victims. 'PredPol' is a predictive policing application used in the UK that requires giving the private company PredPol access to the predictions. Given the sensitive nature of crime data that is outsourced to a private company, the application has been heavily criticised (Hardyns & Rummens, 2018).

Whilst AI offers necessary solutions to the overload of data available online, such as assisting officers in the triage and prioritisation of certain data or highlighting unseen patterns, GCHQ has pointed out that the AI models are not sophisticated enough yet to independently take decisions in the sphere of predictive policing (GCHQ, n.d.). Moreover, these opportunities can also constitute threats. As such, unseen patterns could wrongfully point towards causation, whereas a mere correlation exists. Moreover, existing biases and discrimination in the way law enforcement investigates crime can be reinforced when this data is inserted in the predictive policing models (Amnesty International & Access Now, 2018; Committee of Experts on Internet Intermediaries, 2018; European Union Agency for Fundamental Rights, 2020). Given that predictive policing occurs during the investigation period and not the trial period, the Convention's right to a fair trial does not apply to this pre-trial stage. The application of fair trial rights is triggered by the charges resting on a person and applicable to the subsequent criminal investigation. Consequently, the principle of presumption of innocence is not applicable to the initial stage during which predictive policing techniques are used. However, and agreeing with Blount, it seems inappropriate to dissociate the policing and trial stage when it comes to the presumption of innocence, given the impact this principle has on the entire procedure (Blount, 2021).

4.2.3 Facial Recognition Combined with Surveillance Practices

Another and more intrusive technology used by states in combination with mass surveillance is the use of facial recognition technology (Access Now, 2018). The Council of Europe's Consultative Committee of the Convention for the Protection of Individuals with regard to Automatic Processing of Personal Data (Committee of Convention 108) defines facial recognition technologies as the 'automatic processing of digital images containing individuals' faces for identification or verification of those individuals by using face templates' (Committee of Convention 108, 2021, p. 3). The importance of an individual's facial image has been recognised by the European Court of Human Rights as constituting 'one of the chief attributes of his or her personality, as it reveals the person's unique characteristics and distinguishes the person from his or her peers' (Lopez Ribalda and others v. Spain, 2019, para. 89). As the Committee of Convention 108 points out, facial recognition technology can be used for purposes of verification and identification. Verification, also referred to as 'one-to-one matching', occurs when 'two biometric templates are compared to determine if the person shown on the two images is the same person' (European Union Agency for Fundamental Rights, 2020, p. 7). Verification often occurs at a country's borders when passports are checked through automated border control systems. Identification, or 'one-to-many matching', occurs when 'the template of a person's facial image is compared to many other templates stored in a database to find out if his or her image is stored there' (European Union Agency for Fundamental Rights, 2020, p. 7). The technology will then present a score of likelihood that the person whose facial image is verified and a stored image in the database are one and the same person. Contrary to identification where no external database is consulted, verification will compare the person's facial image to an external database (European Union Agency for Fundamental Rights, 2020).

Whilst the use of facial recognition technology can be of immense value for finding missing children or detecting fraud or identity theft (European Union Agency for Fundamental Rights, 2020; Lazarus et al., 2021), many human rights concerns rise about the quality and the state's use of this technology (European Union Agency for Fundamental Rights, 2020; Mijatovic, 2018). As such, the mere recourse by the state to facial recognition technology constitutes a heavy interference with citizens' right to privacy (Committee of Convention 108, 2021). Alarmingly, companies and states have already recognised the danger of facial recognition technology, and more specifically regarding biases and discrimination that are inherent to such AI-enabled technology. As such, three American cities have banned the use of such technology for policing purposes because of bias concerns. Moreover, Axon, a supplier of body-cameras for law enforcement agencies has announced it would refrain from incorporating any facial recognition technology in its products because of its inherent biases and ethical and discriminatory concerns (Crawford, 2019; Feldstein, 2019).

Although the technology in facial recognition systems has improved significantly, quality concerns still remain. Low quality of images, because of a person's hair or skin colour, the background or illumination, that are stored in the databases used to identify persons may lead to inaccurate results, errors, false positives and false negatives. Research has shown that persons with a darker skin colour will more often be falsely identified or matched with images stored in databases. Consequently, persons with darker skin are more often erroneously accused of crimes they have not committed (Access Now, 2018; European Union Agency for Fundamental Rights, 2020; Feldstein, 2019). If law enforcement officers decide to stop a person solely based on such facial recognition technology that disproportionately falsely identifies persons with darker skin as a match, that person's right to a fair trial will be endangered (European Union Agency for Fundamental Rights, 2020). Furthermore, the person could argue a violation of the prohibition of discrimination given that the decision was made on the basis of the person's skin colour, a protected characteristic under article 10 of the European Convention on Human Rights. Moreover, as with surveillance in general, the recourse by law enforcement agencies to facial recognition technology can lead to a chilling effect on citizens' other fundamental rights, such as the freedom of expression and of assembly, as the citizens might feel discouraged to actively participate in democracy by, for example, attending protests (Access Now, 2018; European Union Agency for Fundamental Rights, 2020).

Live Facial Recognition Technology is the live version of identification through facial recognition technology as it extracts faces from video footage to compare 'them against the facial images in the reference database to identify whether the person on the video footage is in the database of images' (European Union Agency for Fundamental Rights, 2020, p. 8). This technology is worrisome in light of citizens' fundamental right to privacy given that they can constantly be watched. The United Kingdom has been one of the countries that tested this technology in street cameras (European Union Agency for Fundamental Rights, 2020). The European Court of Human Rights has ruled that the right to private life exceeds the private household and can also apply in the public sphere. As such, it ruled that 'a person's reasonable expectation to privacy may be a significant, although not necessarily conclusive, factor' in the assessment of a privacy breach in the public sphere (Lopez Ribalda and others v. Spain, 2019, para. 88). Whilst this right to privacy is not absolute, any state's interference by their use of facial recognition technology will have to stand the 'legality-legitimacy-necessity test'. In light of the necessity requirement, it is questionable whether the use of live facial recognition technology fulfils this requirement, given the grave interference this technology constitutes.

At the level of the United Kingdom, the Court of Appeal of England and Wales ruled in *R (Bridges) v Chief Constable of South Wales Police* [2020] EWCA Civ 1058 that the use of automated facial recognition technology by law enforcement agencies in South Wales violated article 8 of the ECHR. At the level of the Council of Europe, the Committee of Convention 108 has already issued guidelines on the use of facial recognition technology, calling on its member states to respect their citizens' human rights and fundamental freedoms when deploying such technology (Committee of Convention 108, 2021). However, it remains to be seen whether the European Court of Human Rights will be seized with the question and how it will rule on it.

4.3 Drones and Lethal Autonomous Weapons

Unmanned Aerial Vehicles, otherwise known as drones, and lethal autonomous weapons can also be deployed for surveillance purposes, in which case the above-mentioned threats and concerns equally apply to the recourse to this type of AI. However, this technology also constitutes a threat to human rights in itself. Drones and lethal autonomous weapons, the latter understood as 'weapons that once launched select and [destroy] targets without further human intervention' (World Commission on the Ethics of Scientific Knowledge and Technology, 2017, p. 25), can operate on AI that allows them to carry out missions by selecting and attacking targets on their own, without human intervention (World Commission on the Ethics of Scientific Knowledge and Technology, 2017). On the one hand, the recourse by states to such technology raises human rights questions regarding the state's obligation to protect their citizens' lives and questions of legality under international humanitarian law. On the other hand, it raises ethical questions on whether the capability and power to kill should be left with machines (Cummings, 2017; World Commission on the Ethics of Scientific Knowledge and Technology, 2017).

The obligation to protect the life of persons in their jurisdiction and to protect them against intentional deprivation of life, as legally enshrined in article 2 of the ECHR, can conflict with the recourse by states to drones and unmanned aerial vehicles when used in their own jurisdiction or against their own citizens. It is also conceivable that this right would come under tension when foreign states use drones or unmanned aerial vehicles on a states' territory when this recourse leads to physical damage of citizens. The European Court of Human Rights ruled that 'article 2 must be interpreted in so far as possible in light of the general principles of international law, including the rules of international humanitarian law' (Varnava and others v. Turkey, 2009, para. 185). Moreover, the Court found that 'even in situations of international armed conflict, the safeguards under the Convention continue to apply, albeit interpreted against the background of the provisions of international humanitarian law' (Hassan v. the United Kingdom, 2014, para. 104). Consequently, a combined analysis of human rights law under the European Convention on Human Rights with the principles of international humanitarian law is required.

From an international humanitarian law perspective, AI-enabled weapons raise legal questions, given that they can be deployed to orchestrate targeted attacks. Under article 49(1) of the Protocol Additional to the Geneva Conventions of 12 August 1949 and relating to the Protection of Victims of International Armed Conflicts 1977 (AP I), targeted attacks are to be understood as 'acts of violence against the adversary, whether in offence or in defence' that can be directed towards objects (targeted attacks) or persons (targeted killings). For targeted attacks to be legal, three major principles are to be respected: the principles of distinction, proportionality (AP I, art. 51(5)(b)) and precaution (AP I, art. 57). Whilst the principle of precaution, which requires that 'the attacking party must still take all feasible precautions to choose means and methods of warfare that will avoid as much incidental harm to civilians as possible' (Melzer, 2019, p. 102), even if the attack has been deemed proportionate, is less relevant to the ethical and legal analysis of the deployment of AI-enabled lethal autonomous weapons, the former two principles require a more thorough analysis in light of international humanitarian law.[6]

The principle of distinction restricts the objects or persons of attack as it prohibits targeting civilians or civilian objects (AP I, art. 51(2)) unless the civilians take a direct part in hostilities (AP I, art. 51(3)). Since civilians and civilian objects cannot be targeted if they do not take a direct part in hostilities, the military forces of a party to the conflict must distinguish this category of persons and objects from legitimate targets, being military objectives and combatants (AP I, art. 48). However, it is questionable whether an AI-operated machine will be able to distinguish civilians from military personnel (Szocik & Jurkowska-Gomułka, 2021). Moreover, as the World Commission on the Ethics of Scientific Knowledge and Technology rightly points out, it is also questionable whether AI-enabled drones and lethal autonomous weapons will notice a person's changed status (World Commission on the Ethics of Scientific Knowledge and Technology, 2017), such as a civilian deciding to take part in the hostilities, or a civilian who previously took part in the hostilities, but withdraws from it. It is highly doubtful whether machines will be able to correctly apply the principle of distinction. The second principle, the principle of proportionality requires that the direct and concrete military advantage of a lawful targeted attack outweighs the suffering and incidental damages (Melzer, 2019). The targeted attack cannot serve 'political, economic or other non-military

[6] Due to the limited scope of this work, the different principles and the international humanitarian law in general will not be discussed in depth. For a more detailed analysis of these topics, see: Melzer (2019).

benefits' (Melzer, 2019, p. 101). Assessing the military advantage against the potential damages is an inherently nuanced analysis that requires a human approach (World Commission on the Ethics of Scientific Knowledge and Technology, 2017). Machines do not have the human brain's capacity to take into account nuance and circumstances (Access Now, 2018; World Commission on the Ethics of Scientific Knowledge and Technology, 2017). Ethically, the proportionality assessment of a targeted attack cannot be left to machines but requires human intervention.

Other ethical questions also arise when taking into account that drones and lethal autonomous weapons can, like any AI-operated machine, be spoofed or hacked (World Commission on the Ethics of Scientific Knowledge and Technology, 2017). The spoofing or hacking of such machines could lead to disastrous consequences, such as non-state actors taking over control of drones or lethal autonomous weapons in the middle of an operation and diverting these machines towards other illegitimate targets. Consequently, these AI-enabled machines could end up killing civilians or destroying civilian targets to the advantage of non-state actors, which could trigger a state's responsibility for having violated its citizens' fundamental right to life as protected under article 2 of the European Convention on Human Rights. When granting machines, such as drones and lethal autonomous systems, the power to kill, one should always bear in mind that control over such machines can be lost. The loss of control over such powerful technology can have disastrous consequences. Moreover, when drones and lethal autonomous weapons are utilised by non-state actors to perpetrate attacks, politically attributing these attacks might become difficult and questions of accountability rise (Access Now, 2018; Johnson, 2019). Consequently, the World Commission on the Ethics of Scientific Knowledge and Technology recommended reconsidering the recourse to such AI-enabled machines, as has been done in the past regarding biological and chemical weapons and anti-personnel mines (World Commission on the Ethics of Scientific Knowledge and Technology, 2017).

4.4 AI-Enabled Foreign State Disinformation

When foreign states interfere in a state's political scene by online disinformation campaigns, this latter state's national security is endangered. The UK's definition of national security explicitly covers 'actions intended to overthrow or undermine parliamentary democracy by political (…) means' (Security Service Act 1989, s.1(2)). The UK's GCHQ has highlighted this threat, stating that AI can be used by hostile actors to orchestrate disinformation campaigns and attacks 'by automating the production of false content to undermine public debate' (GCHQ, n.d.). AI can also be deployed to automate political targeting tailored to online individual profiles (GCHQ, n.d.). Besides the political threat foreign state disinformation constitutes to democracies, such campaigns can also violate citizens' right to free elections and their absolute right to hold opinions (Committee of Experts on Internet

Intermediaries, 2018; United Nations Special Rapporteur on the promotion and protection of the right to freedom of opinion and expression, 2018).

The right to free elections, as protected under article 3 of Protocol No. 1, has been recognised to be inherently linked to the enjoyment of the right to freedom of expression since it covers the free circulation of all kinds of opinions and information in pre-election phase (Bowman v. the United Kingdom, 1998; Mathieu-Mohin and Clerfayt v. Belgium, 1987). This is especially relevant in an era where political state disinformation campaigns interfere with democratic elections to stir the public opinion into a certain direction, such as was the case with the Russian interference in the 2016 US Presidential elections (Special Counsel Robert S. Mueller, 2019). The recourse by foreign states to AI-enabled online disinformation campaigns consequently violates citizens' fundamental right to free elections. Moreover, AI-enabled disinformation campaigns interfere with the information a person views (especially on social media), consequently violating their right to freedom of opinion. This illegal interference with the right to free elections and to freedom of opinion is further amplified by the way social media platforms, where such disinformation is spread, function and how this can be abused. This amplification can occur in several ways, including by automation, paid advertisement and recommendation systems.

First, the online social media platforms rely on automation, which is a system that puts content similar to what a user has previously engaged with on that person's newsfeed. Therefore, the AI models that analyse the previously watched content play an important role in what the viewer will or will not see next (Llanso et al., 2020). When a person has been watching several videos of funny animals, the automation system will fill that user's newsfeed with similar animal videos (Dieu et al., 2021). When a person's newsfeed has been targeted by foreign disinformation content, the person's newsfeed will contain even more similar content. Moreover, increasing the active engagement of the users on the platform is partly based on matching these users with content they find interesting. This content can be uploaded by other users, who are sometimes willing to pay to appear predominantly on the user's newsfeed to increase their visibility (Access Now, 2018; Elkin-Koren, 2020). Hence, using these paid advertisement options allows malicious foreign states to optimise their visibility to targeted individuals.

The recommendation system of social media platforms is linked to this automation. This AI-enabled system recommends certain content to a viewer based on previously watched content. A user's newsfeed is thus not neutral or random and will only magnify the previously seen opinion (Elkin-Koren, 2020). This system prevents a person from seeing different opinions regarding a specific topic, unless when specifically looking for the opposite opinion. Because of the threat to further indoctrination by, amongst others, foreign disinformation, states and private social media platforms have an interest in combating this spread by detecting these online threats (Dieu et al., 2021).

4.5 Combating Illegal Content Online: Detection of Threats

Private social media platforms and governments collaborate to detect online threats, such as when tackling terrorist propaganda or child sexual exploitation material. Taking the example of terrorism, which constitutes a national security threat, the illimited reach terrorist groups have online and use for propaganda, recruitment and mobilisation (Weimann, 2004) justifies this intervention by the state (Dieu et al., 2021). When cooperating with private social media platforms, states can intervene at two stages. First, governments use monitoring programmes that allow these governments to 'collect social media information and feed it to AI-powered programs to detect alleged threats' (Access Now, 2018, p. 20). The state's Internet Referral Units flag content deemed illegal to these platforms. After content has been flagged by these units, the social media platform (generally) takes that content down (Chang, 2018; Ellerman, 2016). Second, governments increasingly pressure social media platforms to use automation mechanisms to expeditiously take down content that violates the platform's terms and conditions.

Given the enormous amount of data and content on those platforms, it would be unrealistic to require those platforms to analyse every referred content's legality manually. Social media platforms are consequently required to use AI to be able to comply with the government's requirements and pressure (Dieu et al., 2021). Hence, for the protection of national security, states indirectly utilise AI to diminish the online presence of illicit content. However, AI does not have the human eye and brain to interpret the context or subtleties of online content (Cambron, 2019; Huszti-Orban, 2018; United Nations Special Rapporteur on the promotion and protection of the right to freedom of opinion and expression, 2018). Whilst companies, such as Google and Microsoft, are increasingly turning to machine learning to train their AI to better understand subtleties of particular contexts such as humour or satire (Gollatz et al., 2018), excessive takedowns of legitimate content are still problematic. An example of such an excessive takedown that disregarded the context concerns Al-Mutez Billah's YouTube channel that was considered the 'digital archive of the Syrian war'. Due to YouTube's automatic takedown procedure, his channel's content that collected legitimate evidence of the ongoing Syrian war was taken down (Access Now, 2018; TRTWorld, 2021; Qantara, 2021). Hence, the content moderation mechanisms can take down legitimate speech, which is a violation of the users' right to freedom of expression (Mijatovic, 2018).

Excessive takedowns of legitimate speech are not the only type of errors and concerns that can be raised when relying on AI. As such, content moderation can have a chilling effect on citizens who want to express legitimate speech. These citizens might be less inclined to exercise their right to freedom of expression on platforms that are known to silence illegitimate, but also legitimate speech (Access Now, 2018). Moreover, false positives, understood in this context as legitimate online content erroneously considered as illegal content, and false negatives, being illegal content that is not recognised as such and remains on the platform, in the filtered contents also endanger the users' right to freedom of expression (Dieu et al.,

2021; Fernandez & Alani, 2021; Jorgensen & Pedersen, 2017). On a small scale, these errors might seem insignificant, but once scaled to the enormous amount of content daily shared on social media platforms, the false positive and negative margins become enormous (Committee of Ministers of the Council of Europe, 2020; European Union Agency for Fundamental Rights, 2020). Overall, the states' recourse to AI for the prevention of further dissemination of illegal content that threatens national security is a positive application, but the negative consequences on citizens' right to freedom of expression should be borne in mind.

5 Conclusion

This work examined the legal, human rights and ethical concerns related to the recourse by states to artificial intelligence for national security purposes. Whilst the enormous amount of online data requires states to use AI, the increasing recourse to AI for national security purposes is alarming. The technology in AI is inherently biased, contains errors and leads to false positives and false negatives. When such technology is then developed for purposes of national security, several human rights of citizens subjected to such technology are endangered.

First, the extensive and increasing incorporation of AI in surveillance practices constitutes a threat to citizens' human rights. As such, the recourse by states to AI for national security purposes is often tacitly extended to public safety purposes. This unchallenged broadening of the initial scope threatens the subjected citizens' right to privacy and might endanger their right to freedom of assembly due to the chilling effect of surveillance. The deployment of surveillance for predictive policing purposes threatens citizens' right to a fair trial given the inherent biases and errors AI contains. Last, the combined use of facial recognition technology with mass surveillance practices menaces the citizens' right to privacy, right to fair trial and right to equality. Second, using AI in drones and lethal autonomous weapons endangers the right to life of citizens, creates challenges of legality under international humanitarian law and raises ethical concerns, such as whether machines should be given the power to kill. Third, the development of AI in disinformation campaigns that foreign states channel through social media technology that amplifies that discourse is highly worrisome in light of the social media platform's users' right to free elections and freedom of opinion. Last, whilst the recourse to AI by states to counter online illegal content should overall be encouraged, it still constitutes a violation of a person's right to free speech when legitimate content is taken down. The chilling effect such takedowns generates is worrisome. It is highly unlikely states will in the future withdraw or refrain from using AI for national security purposes. However, with this further incorporation of AI, governments, legislators, national assemblies and the general public should become more aware of the dangers such incorporation creates. A human rights–centric approach is therefore required when states deploy AI for national security purposes.

References

Access Now. (2018). *Human rights in the age of artificial intelligence*. https://www.accessnow.org/cms/assets/uploads/2018/11/AI-and-Human-Rights.pdf

Amnesty International & Access Now. (2018). *The Toronto Declaration: Protecting the right to equality and non-discrimination in machine learning systems*. https://www.accessnow.org/cms/assets/uploads/2018/08/The-Toronto-Declaration_ENG_08-2018.pdf

Babuta, A., Oswald, M., & Janjeva, A. (2020). *Artificial intelligence and UK National Security: Policy considerations*. Royal United Services Institute for Defence and Security Studies. https://static.rusi.org/ai_national_security_final_web_version.pdf

Blount, K. (2021). Applying the presumption of innocence to policing with AI. Artificial intelligence, big data and automated decision-making in criminal justice. *Revue Internationale de Droit Pénale, 92*(1), 33–48. https://orbilu.uni.lu/bitstream/10993/48564/1/Blount%20RIDP%20PDF.pdf

Bovens, G. R., & Schillemans, T. (2016). *The Oxford handbook of public accountability*. Oxford University Press.

Bychawska-Siniarska, D. (2017). *Protecting the right to freedom of expression under the European convention on human rights: A handbook for legal practitioners*. Council of Europe. https://rm.coe.int/handbook-freedom-of-expression-eng/1680732814

Cambridge Dictionary. (n.d.). Algorithm. In Dictionary.cambridge.org. Retrieved March 12, 2022, from https://dictionary.cambridge.org/dictionary/english/algorithm

Cambron, R. J. (2019). World war web: Rethinking "aiding and abetting" in the social media age. *Case Western Reserve Journal of International Law, 51*(1), 293–325.

Chang, B. (2018). From internet referral units to international agreements: Censorship of the internet by the UK and EU. *Columbia HR Law Review, 49*(2), 114–212.

Consultative Committee of the Convention for the Protection of Individuals with regard to Automatic Processing of Personal Data. (2021). *Guidelines on facial recognition*. Council of Europe. https://rm.coe.int/guidelines-on-facial-recognition/1680a134f3

Committee of Experts on Internet Intermediaries. (2018). *Algorithms and human rights: Study on the human rights dimensions of automated data processing techniques and possible regulatory implications*. Council of Europe. https://edoc.coe.int/en/internet/7589-algorithms-and-human-rights-study-on-the-human-rights-dimensions-of-automated-data-processing-techniques-and-possible-regulatory-implications.html

Committee of Ministers of the Council of Europe. (2020). *Recommendation CM/Rec(2020)1 of the Committee of Ministers to member States on the human rights impacts of algorithmic systems*. Council of Europe. https://search.coe.int/cm/pages/result_details.aspx?objectid=09000016809e1154

Council of Europe. (n.d.). *Some definitions*. Retrieved March 11, 2022, from https://www.coe.int/en/web/echr-toolkit/definitions

Crawford, K. (2019, August 27). *Halt the use of facial-recognition technology until it is regulated*. Nature. https://www.nature.com/articles/d41586-019-02514-7

Cummings, L. M. (2017). *Artificial intelligence and the future of warfare*. Chatham House. https://www.chathamhouse.org/sites/default/files/publications/research/2017-01-26-artificial-intelligence-future-warfare-cummings-final.pdf

Da Silva, J. R. (2017). *Jihadist terrorism and EU responses – Current and future challenges*. Institut für Europa – und Sicherheitspolitik. https://www.aies.at/download/2017/AIES-Fokus%2D%2D2017-06.pdf

Dieu, O., Dau, P. M., & Vermeulen, G. (2021). *ISIL terrorists and the use of social media platforms. Are offensive and proactive cyber-attacks the solution to the online presence of ISIL?* [Master Thesis, University of Ghent]. Lib UGent https://libstore.ugent.be/fulltxt/RUG01/003/007/897/RUG01-003007897_2021_0001_AC.pdf

Elkin-Koren, N. (2020). Contesting algorithms: Restoring the public interest in content filtering by artificial intelligence. *Big Data & Society, 7*(2), 1–13. https://journals.sagepub.com/doi/full/10.1177/2053951720932296

Ellerman, J. (2016). Terror won't kill the privacy star. Tackling terrorism propaganda online in a data protection compliant manner. *ERA Forum, 17*(4), 555–582.

Enarsson, T., Enqvist, L., & Naarttijarvi, M. (2022). Approaching the human in the loop – Legal perspectives on hybrid human/algorithmic decision-making in three contexts. *Information & Communications Technology Law, 31*(1), 123–153. https://doi.org/10.1080/13600834.2021.1958860

European Court of Human Rights. (2013). *National Security and European case-law*. Research Division. Council of Europe. https://rm.coe.int/168067d214

European Court of Human Rights. (2021a). *Guide on article 2 of the European convention on human rights: Right to life*. Council of Europe. https://www.echr.coe.int/Documents/Guide_Art_2_ENG.pdf

European Court of Human Rights. (2021b). *Guide on article 6 of the European convention on human rights: Right to a fair trial (criminal limb)*. Council of Europe https://www.echr.coe.int/documents/guide_art_6_criminal_eng.pdf

European Court of Human Rights. (2021c). *Guide on article 11 of the European convention on human rights: Freedom of assembly and association*. Council of Europe. https://www.echr.coe.int/Documents/Guide_Art_11_ENG.pdf

European Court of Human Rights. (2021d). *Guide on article 14 of the European convention on human rights and on article 1 of protocol no. 12 to the convention: Prohibition of discrimination*. Council of Europe. https://www.echr.coe.int/Documents/Guide_Art_14_Art_1_Protocol_12_ENG.pdf

European Court of Human Rights. (2022). *Mass surveillance*. Council of Europe. https://www.echr.coe.int/documents/fs_mass_surveillance_eng.pdf

European Union Agency for Fundamental Rights. (2020). *Facial recognition technology: Fundamental rights considerations in the context of law enforcement*. https://fra.europa.eu/sites/default/files/fra_uploads/fra-2019-facial-recognition-technology-focus-paper-1_en.pdf

Feldstein, S. (2019). *The global expansion of AI surveillance*. Carnegie Endowment for International Peace. https://carnegieendowment.org/2019/09/17/global-expansion-of-ai-surveillance-pub-79847

Fernandez, M., & Alani, H. (2021). Artificial intelligence and online extremism: Challenges and opportunities. In J. McDaniel & K. Pease (Eds.), *Predictive policing and artificial intelligence* (pp. 132–162). Routledge. http://oro.open.ac.uk/69799/1/Fernandez_Alani_final_pdf.pdf

GCHQ. (n.d.). *Pioneering a new National Security: The ethics of artificial intelligence*. Retrieved March 1, 2022, from https://www.gchq.gov.uk/artificial-intelligence/accessible-version.html

Gollatz, K., Beer, F., & Katzenbach, C. (2018). The turn to artificial intelligence in governing communication online (HIIG workshop report). *Big Data & Society* (special issue), https://www.ssoar.info/ssoar/bitstream/handle/document/59528/ssoar-2018-gollatz_et_al-The_Turn_to_Artificial_Intelligence.pdf?sequence=1&isAllowed=y&lnkname=ssoar-2018-gollatz_et_al-The_Turn_to_Artificial_Intelligence.pdf

Haney, B. S. (2020). Applied artificial intelligence in modern warfare and National Security Policy. *Hastings Science and Technology Law Journal, 11*(1), 61–100. https://repository.uchastings.edu/hastings_science_technology_law_journal/vol11/iss1/5

Hardyns, W., & Rummens, A. (2018). Predictive policing as a new tool for law enforcement? Recent developments and challenges. *European Journal on Criminal Policy and Research, 24*, 201–218. https://doi.org/10.1007/s10610-017-9361-2

Heilemann, J. (2021). Click, collect and calculate: The growing importance of big data in predicting future criminal behaviour. Artificial intelligence, big data and automated decision-making in criminal justice. *Revue Internationale de Droit Pénale, 92*(1), 49–67. http://real.mtak.hu/133496/1/RIDP_2021_1_Karsai.pdf

Huszti-Orban, K. (2018). Internet intermediaries and counter-terrorism: Between self-regulation and outsourcing law enforcement. In T. Minarik, L. Lindstrom, & R. Jakschis (Eds.), *10th international conference on cyber conflict: CyCon X: Maximising effects* (pp. 227–243). NATO CCD COE Publications. https://doi.org/10.23919/CYCON.2018.8405019

Information Commissioner's Office. (n.d.). *National security and defence*. Retrieved March 7, 2022, from https://ico.org.uk/for-organisations/guide-to-data-protection/guide-to-the-general-data-protection-regulation-gdpr/national-security-and-defence/

Johnson, J. (2019). Artificial intelligence & future warfare: Implications for international security. *Defense & Security Analysis, 35*(2), 147–169. https://doi.org/10.1080/14751798.2019.1600800

Jorgensen, R. F. (2019). *Human rights in the age of platforms*. The MIT Press.

Jorgensen, R. F., & Pedersen, A. M. (2017). Online service providers as human rights arbiters. In M. Taddeo & L. Floridi (Eds.), *Law, governance and technology series: The responsibilities of online service providers* (Vol. 31, pp. 179–199). Springer.

Koehler-Schindler, M.. (2021, September 10). *20 Years after 9/11: Why we must protect human rights while countering terrorism*. European Leadership Network. https://www.europeanleadershipnetwork.org/commentary/20-years-after-9-11-why-we-must-protect-human-rights-while-countering-terrorism/

Koops, B.-J. (2021). The concept of function creep. *Law, Innovation and Technology, 13*(1), 29–56. https://doi.org/10.1080/17579961.2021.1898299

Lazarus, L., Le Toquin, J.-C., Magri o Aires, M., Nunes, F., Staciwa, K., Vermeulen, G., & Walden, I. (2021). *Respecting human rights and the rule of law when using automated technology to detect online child sexual exploitation and abuse* (Independent experts' report). Directorate General of Human Rights and Rule of Law & Directorate General of Democracy. https://rm.coe.int/respecting-human-rights-and-the-rule-of-law-when-using-automated-techn/1680a2f5ee

Leslie, D. (2019). *Understanding artificial intelligence ethics and safety: A guide for the responsible design and implementation of AI systems in the public sector*. The Alan Turing Institute. https://doi.org/10.5281/zenodo.3240529

Llanso, E., Van Hoboken, J., Leerssen, P., & Harambam, J. (2020). *Artificial intelligence, content moderation, and freedom of expression*. Transatlantic Working Group. https://www.ivir.nl/publicaties/download/AI-Llanso-Van-Hoboken-Feb-2020.pdf

Loideain, N. N. (2019). A bridge too far? The investigatory powers act 2016 and human rights law. In L. Edwards (Ed.), *Law, policy, and the internet* (pp. 165–192). Hart Publishing.

Lord Phillips of Worth Matravers. (2010, June 8). *The Challenges of the New Supreme Court*. Gresham College. https://www.gresham.ac.uk/lectures-and-events/the-challenges-of-the-new-supreme-court

McCarthy, J. (n.d.). *What is AI? / Basic Questions*. Retrieved March 7, 2022, from http://jmc.stanford.edu/artificial-intelligence/what-is-ai/index.html

McKendrick, K. (2019). *Artificial intelligence prediction and counterterrorism*. Chatham House. https://www.chathamhouse.org/sites/default/files/2019-08-07-AICounterterrorism.pdf

Melzer, N. (2019). *International humanitarian law: A comprehensive introduction*. ICRC.

Mijatovic, D. (2018, July 3). *In the era of artificial intelligence: Safeguarding human rights*. Open Democracy. https://www.opendemocracy.net/en/digitaliberties/in-era-of-artificial-intelligence-safeguarding-human-rights/

Moses, L. B., & Janet, C. (2018). Algorithmic prediction in policing: Assumptions, evaluation, and accountability. *Policing and Society, 28*(7), 806–822.

Osoba, O. A., & Welser IV, W. (2017). *The risks of artificial intelligence to security and the future of work*. RAND Corporation. https://www.rand.org/content/dam/rand/pubs/perspectives/PE200/PE237/RAND_PE237.pdf

Oswald, M. (2020, May 13). *AI and national security: Learn from the machine, but don't let it take decisions*. AboutIntel: https://aboutintel.eu/ai-uk-national-security/

Qantara. (2021, March 8). *Activists in race to save digital trace of Syria war*. https://en.qantara.de/content/activists-in-race-to-save-digital-trace-of-syria-war

Ratcliffe, J. (2014). What is the future… of predictive policing? *Translational Criminology, 6*, 4–5. https://www.academia.edu/26606550/What_Is_the_Future_of_Predictive_Policing

Rodrigues, R. (2020). Legal and human rights issues of AI: Gaps, challenges and vulnerabilities. *Journal of Responsible Technology, 4*. https://doi.org/10.1016/j.jrt.2020.100005

Schwartz, M. (2014). *Challenges at the nexus of security and the promotion and protection of human rights*. Global Center on Cooperative Security. https://www.globalcenter.org/wp-content/uploads/2014/11/14Oct_Security-and-Human-Rights_-Workshop-I-Discussion-Working-Paper_FINAL..pdf

Secretary-General of the United Nations. (2018, August 29). *Note by the Secretary-General on the promotion and protection of the right to freedom of opinion and expression*. United Nations. https://digitallibrary.un.org/record/1643488?ln=en

Special Counsel Robert S. Mueller. (2019). *Report on the investigation into Russian interference in the 2016 presidential election*. U.S. Department of Justice. https://www.justice.gov/archives/sco/file/1373816/download

Surden, H. (2014). Machine learning and law. *Washington Law Review, 89*(1), 87–115. https://scholar.law.colorado.edu/articles/81/

Szocik, K., & Jurkowska-Gomułka, A. (2021). Ethical, legal and political challenges of artificial intelligence: Law as a response to AI-related threats and hopes. *World Futures, 1–17*. https://doi.org/10.1080/02604027.2021.2012876

TRTWorld. (2021, March 8). *Activists accuse YouTube of destroying digital evidence of Syria war*. https://www.trtworld.com/life/activists-accuse-youtube-of-destroying-digital-evidence-of-syria-war-44809

United Nations Office of the High Commissioner for Human Rights (OHCHR). (2008). *Fact sheet no. 32, human rights, terrorism and counter-terrorism*. https://www.refworld.org/docid/48733ebc2.html

United Nations Special Rapporteur on the promotion and protection of the right to freedom of opinion and expression. (2018, August 29). *Promotion and protection of the right to freedom of opinion and expression* (A/73/348). https://digitallibrary.un.org/record/1643488?ln=en#record-files-collapse-header

United Nations Special Rapporteur on the promotion and protection of freedom of opinion and expression. (2019). *Surveillance and human rights* (A/HRC/41/35). https://digitallibrary.un.org/record/3814512?ln=en

Weimann, G. (2004). *How modern terrorism uses the internet*. United States Institute of Peace. https://www.usip.org/sites/default/files/sr116.pdf

World Commission on the Ethics of Scientific Knowledge and Technology. (2017). *Report of COMEST on robotics ethics* (SHS/YES/COMEST-10/17/2 REV.). https://unesdoc.unesco.org/ark:/48223/pf0000253952

Yu, S., & Carroll, F. (2021). Implications of AI in National Security: Understanding the security issues and ethical challenges. In R. Montasari & H. Jahankhani (Eds.), *Artificial intelligence in cyber security: Impact and implications. Security challenges, technical and ethical issues, forensic investigative challenges* (pp. 157–175). Springer. https://doi.org/10.1007/978-3-030-88040-8

The Use of AI in Managing Big Data Analysis Demands: Status and Future Directions

Vinden Wylde, Edmond Prakash, Chaminda Hewage, and Jon Platts

Abstract In the modern age, the context of health, energy, environment, climate crisis, and global Covid-19 pandemic, managing Big Data demands via Sustainable Development Goals and disease mitigation supported by Artificial Intelligence, present significant challenges for a given territory or national boundaries' policies, legal systems, energy infrastructure, societal cohesion, internal and external national security. We look at policy, technical, and legal applications alongside ramifications of relevant policies and practices to highlight key challenges from a global and societal context. This review contributes to developing further awareness of the complexity and demands on policy and technology. In the long term due to these significant changes, inferences of multifaceted policy and data acquisition could present additional compounding challenges regarding civil liberties, data privacy law, and equitable health outcomes, whilst implementing continually evolving policies, practices, and techniques that can weaken infrastructure and present cyber-attack vulnerabilities. As a consequence of local, regional, and international paradigm shifts, Blockchain and Smart Contracts are suggested as part of a solution in providing data protection, transparency, and validity with transactional data to enable further trust between private and public sectors during times of crisis and technological transition processes.

Keywords Blockchain · AI · Big Data · Health · Sustainability · IoT

1 Introduction

Through the Internet of Things (IoT) and its evolution for both private and business environments, data has become an important commodity that provides the backbone to an array of innovative data-driven applications across multiple domains and sec-

V. Wylde (✉) · E. Prakash · C. Hewage · J. Platts
Cardiff School of Technologies, Cardiff Metropolitan University, Cardiff, UK
e-mail: vwylde@cardiffmet.ac.uk; eprakash@cardiffmet.ac.uk; chewage@cardiffmet.ac.uk; jplatts@cardiffmet.ac.uk

R. Montasari (ed.), *Artificial Intelligence and National Security*,
https://doi.org/10.1007/978-3-031-06709-9_3

tors. Domains, such as digital health, production of goods in industry 4.0, and food supply management, all have common use-cases that depend upon (permanently in some cases) the provision for data-driven supply and demand models that are generated by multiple data producers, made available to all consumers, and available at the right specification (i.e., quality and quantity) (Przytarski et al. 2022). Here, SMEs, in particular, remain vulnerable in digital supply chains and ecosystems (Global Cybersecurity Outlook 2022). Moreover, in recent times, the cost of living crisis in the United Kingdom (UK) continues to permeate traditional and digital news media outlets, in particular, how gas, electricity prices, and citizen taxes could significantly rise in the coming months (Energy Bills Could Rise 2021) and how they affect small businesses in the wake of covid-19 (Energy Costs and Covid Pose 'existential threat' to UK's Small Businesses 2021).

Due to a perfect storm of market forces (including wholesale gas prices being five times higher than two years ago, with potential rises of 12% on household bills (The Guardian View on an Energy Price Shock: A Crisis in the Making 2021)), the UK's energy system and economy brings vast uncertainty to its energy suppliers, heavy industry, factories, businesses, and farmers (i.e., consumers may experience empty supermarket shelves due to producing carbon dioxide and dry ice being unprofitable in preserving meat products, ultimately giving rise to the potential of introducing a '*1970's style three-day week*' (The Guardian View on an Energy Price Shock: A Crisis in the Making 2021)), which in turn affects and contributes to a wave of energy suppliers collapsing with UK households footing increasingly unaffordable energy bills, hence producing and shaping the energy crisis (What Caused the UK's Energy Crisis? 2021).

To account for these transactions (i.e., energy suppliers and households), databases are maintained and created in facilitating the management of these vast amounts of data to ensure competitiveness and fairness across business and private sectors. However, with the signing of the Paris Agreement in 2015, 196 countries signed up to fight climate change and agree on measures to mitigate carbon emissions including the setting of the United Nations (UN) 17 Sustainable Development Goals (SDGs).

These complex undertakings with involvement of state (i.e., government, institution, or country) and non-state actors (i.e., Non-Government Organisations [NGO's], individuals: not allied to a state) in mitigation of additional uncertainties and risks utilise Artificial Intelligence (AI) and Big Data (BD) technologies that could provide a way to anticipate the cost of energy through implementing smart meters in a smart grid configuration, for example. Together, these technologies and processes support in combating climate change via energy and carbon trading mechanisms, which ultimately enhances in the additional development of solutions to include health monitoring (i.e., the current covid-19 pandemic).

Here, there are knowledge gaps in the use-cases of Blockchain (BC); its approaches traditionally included systems and methods that utilise Bitcoin and Ethereum (Crypto-currencies) as a substitute for fiat currency, alongside consensus protocols, off-chain storage, anonymous signatures, and non-interactive zero-knowledge proofs. These concepts allow for the provision of validity, anonymity, with transparency, especially when utilised in parallel with organisational and corporate policy deployment, security services and health providers can all function harmoniously in executing domestic and legal activities (Wylde et al. 2022).

However, as the globe recovers from the Covid-19 pandemic, if these innovative technologies are to contribute to development of solutions, then there has to be appropriate methodologies, tools, and standards to validate disruptive technologies such as BC and the hype surrounding the BC generalised approach, towards a more evidence-based narrative (Aysan et al. 2021). Even though BCs potential is promising, in reality the challenges such as not enough BC-backed applications and under-studied social impacts mean that although the potential is there from BC to trigger innovation, the technology itself is yet to mature.

- **Objective of this paper:** The main objective of this paper is to define and describe national and international challenges in AI and BD management policy cohesiveness, to highlight architecture mechanisms to include BC use-cases, to promote methods in utilising more disruptive technologies such as BC in attaining transparency, accountability, and to deliver a more equitable playing field for consumers and suppliers moving into a more sustainable and green economy.
- **Problem Description:** As identified in the literature, BC technologies have a number of application domains across multiple sectors (Aklilu and Ding 2022; Clavin et al. 2020; Kim and Huh 2020; Alladi et al. 2019). Moreover, there are insufficient BC-backed applications and studies in support of these sectors. Main contributions here show that in Przytarski et al. (2022), for example, that between several parties, the need for a robust and trustworthy basis of dialogue to not involve external authorities is preferred. Additional key findings indicate that a single point of truth (i.e., sharing of company data) is necessary to enable transparency (i.e., stakeholder auditability) in that published data is visible to all participants on the BC network alongside data provenance measures (e.g., full historical data availability). In addition, to include the provision for immutable data (i.e., legacy and new data cannot be altered [modified or deleted]) and data to not remain susceptible to attack from within the storage medium.
- **Paper Contributions:** Whilst surveying AI strategies, management of BD and BC mechanisms, our contribution is to highlight the complex array of policies, architectures, and outcomes in providing user and consumer methodologies to partake in an economy that is moving towards zero-carbon emissions. We highlight key terms to enable an equitable, transparent, and auditable experience for consumers and suppliers in moving towards more trusted outcomes. We surveyed work on AI strategies, BD management, and BC related works from IEEE Explore, ACM Digital Libraries, MDPI, and grey literature to include company and government guidance documents.

We looked at previous literature related to AI and BD management strategies with BC use-cases to highlight and support key challenges in policy, technology, and its application.

The chapter structure is as follows.

Section 2: This analyses the literature in terms of the UK AI strategy in relation to energy suppliers and utilisation of AI in national grid resilience, sustainability, and national security. These strategies are then discussed in terms of meeting SDGs and targets of low carbon emissions and for future generations (2030).

Section 3: As a form of data collection to aid services in healthcare surveillance techniques, smart meters are highlighted and discussed to ascertain challenges in relation to technologies such as Internet of Energy (IoE) and SDG comprehensiveness with real-world application in data sharing between suppliers and consumers.

Section 4: Here, we propose utilising BC as a means of data verification, accountability, and transparency in aiding in effectively managing data from a consumer point of view with use-cases.

Section 5.1: As a result of findings, key information is presented to aid in understanding the political, social, environmental, and economic uncertainties, whilst shedding further light upon the scale and depth of initiating significant change from leadership to the disruption and misinformation that remains open to abuse and opportunism.

Section 5.2: This looks at targeted approaches in leadership and governance to include deploying strategies that may present cohesiveness challenges in times of crises. AI strategy funding allocation is encouraged to be prioritised, whilst data protection should be more at the forefront of the public interest with regard to the acquisition of data, to and from state and non-state actors. This is to make transparency and accountability more favourable over competitiveness and global market dominance in future. Partnerships and education alongside more innovative technologies for effective resource deployment such as BC are suggested.

2 Previous Work

2.1 Policy

2.1.1 People: UN Future Generations (2030)

On every continent, every country is continually being affected by climate change. Due to the global crisis, national economies and human lives are being significantly disrupted thus costing communities and countries today and forecast to affect people further into the future. These major disruptions include increasing extreme weather events, changing weather patterns, and the rising of sea levels due to Greenhouse Gas Emissions (GGE), said to be at their highest levels in history, thus driving climate change from increasing human activities.

Without action, the poorest and most vulnerable of society will continue to be affected by the world's average surface temperature, set to rise and surpass 3 °C within this century (Climate Action: Tackling Climate Change 2021). Building upon the Kyoto Protocol of December 1997, which focused around already developed countries in reducing GGEs (Kim and Huh 2020), the Paris Agreement 2015

United Nations
Sustainable Development Goals

(1) No Poverty
(2) Zero Hunger
(3) Good Health and Wellbeing
(4) Quality Education
(5) Gender Equality
(6) Clean Water and Sanitation
(7) Affordable and Clean Energy
(8) Decent Work and Economic Growth
(9) Industry Innovation and Infrastructure
(10) Reduced Inequalities Growth
(11) Sustainable Cities and Communities
(12) Responsible Consumption and Production
(13) Climate Action
(14) Life Below Water
(15) Life on Land
(16) Peace, Justice and Strong Institutions
(17) Partnerships for the Goals

Fig. 1 Sustainable Development Goals. United Nations (2015) Resolution adopted by the General Assembly on 25 September 2015, Transforming our world: the 2030 Agenda for Sustainable Development (A/RES/70/1 Archived 28 November 2020 at the Wayback Machine), https://upload.wikimedia.org/wikipedia/commons/d/d5/N1529189.pdf

recognised that climate change and GGEs is a global challenge that further requires all countries to contribute to the setting of emissions targets. To achieve 'a better future for all', in September 2015, all 193 Member States of the UN set out a path to end extreme poverty, fight inequality, and injustice and to ultimately protect the planet.

Therefore, 17 Sustainable Development Goals (SDG), although all intrinsically linked, provided a 15-year plan and set of objectives to work towards in taking on this immense challenge (Fig. 1).

2.1.2 Energy: COP 26 UN Climate Change Conference

With the recent 26th Conference of Parties (COP 26), from 31 October to 12 November 2021, held in the Scottish Event Campus (SEC), Glasgow UK, the Paris Climate Accords, or '*The Paris Agreement, 2015*', a legally binding international treaty on Climate Change, was adopted by 196 Parties in Paris on the 12th of December 2015, effective from the 4th of November 2016. Its goal, to limit global warming to below 2 °C (ideally, 1.5 °C), is compared to pre-industrial levels. For this long-term goal to be achieved, affiliated countries aim to be climate neutral by mid-century. With regard to goal (13) Climate Action, for example, the call for urgent action in the combat against climate change and its effect upon the globe, provide clear guidance for participating countries in meeting the 2030 agenda for sustainable development through the Paris Agreement 2021, Martins et al. (2022).

Grid Resilience, Sustainability, and National Security The UK is one of the many western democracies implementing significant change to their national grid strategies inclusive of resilience and sustainability. As with the United States of America (USA) on November 30, 2020, the Department of Energy (DOE) formerly announced the creation of a Grid Resilience for National Security (GRNS) subcommittee as part of the Electricity Advisory Committee (EAC), to enhance and support the DOE in modernising the electricity delivery infrastructure, to utilise GRNS guidance to anticipate increasing threats, and to develop new ways in risk management and threat mitigation.

The acting Assistant Secretary of the Office of Electricity stated that '*Our electric grid is vital to national security, public health and safety, and the U.S. economy*' and added that it is a top priority of DOE and the US government in aiding industry and grid resilience strategies together in deepening cooperation efforts in addressing these threats (Department of Energy's Electricity Advisory Committee Establishes the Grid Resilience for National Security Subcommittee 2021).

Later in March 2021, the UK's National Grid produced a holistic (Energy Whole System) report that highlights the interactions between electricity, gas (methane, bio-gas, and hydrogen), and liquid fuels (bio-fuel and oil), and how these energy sources can best facilitate in delivering Net Zero gas emission energy for communications, technology, heat, water, and transport. Stating further that '*the best mix of energy* ***should*** *provide economic, reliable and resilient green energy for UK society*' (Electricity Transmission: Our resilient Whole System approach 2021).

United Kingdom National AI Strategy In September 2021, the UK published a National AI Strategy that focuses on the potential and speed at which AI is growing globally, recognising that with AI comes the ability to change the rules concerning industry, to enhance economic growth and transform many areas of everyday life. On a basic level for administrative burdens, AI can reduce and resolve many challenges to include, allocation in resource management by performing complex tasks, answering questions, and for searching for information whilst filling out documents. However, citizen satisfaction with current government contributions leaves much to be desired (Wirtz et al. 2019).

As the UK is a superpower regarding AI, the strategy boasts in its global leadership [over the next decade], research capabilities, innovation, and wealth of talent in progressive regulatory and business capacities (HM Government National AI Strategy 2021). These successes were supported by the 2017 Industrial Strategy, published in November: the UK governments vision in AI innovation making the UK a global centre, and in April 2018, the UK AI ecosystem attained a £1 Billion deal to further enhance the UK's global position as a leader and developer of AI technology (HM Government National AI Strategy 2021).

This is shown throughout the UK's energy sectors with further plans to change the way health services gather and disseminate data through the national grid (or smart grid [IoE]). As AI is repetitive in nature thus able to build upon and utilise the linguistic, auditory, spatial, and visual aspects inherent in the technology, government agencies are able to actively plan and anticipate legacy systems due to opportunities from AI, although the types of problems that exist where applications of AI are appropriate, range from expert shortages in enhancing large datasets (BD) for fresh insights and outputs and utilising prediction tools to help responding to time-sensitive cases (Wirtz et al. 2019).

2.2 *Technology*

2.2.1 Artificial Intelligence (AI)

Artificial Intelligence (AI) has gained special attention as an inter-disciplinary field of research that continues to benefit societies, economics, the public sectors, and beyond. The application of self-learning algorithms and digitisation significantly changes and advances areas such as the business sector with innovations that justify its relevancy (Wirtz et al. 2019). An example could be in utilising software to monitor worker productivity via desktop, calendar, at times to include the webcam. Even though this tool could be named as 'Productivity Software', this could easily be utilised in exploitation (Positive AI Economic Futures Insight Report November 2021) of workers.

Apart from identifying AI use-cases, it could also be said that the cost implications due to not having the right skills and data infrastructure in place are high (Empowering AI Leadership: AI C-Suite Toolkit 2022). AI is based around symbol manipulation, deductive logic, binary (hard computing method), and statistical methods (Riekki and Mämmelä 2021), although over decades the term AI has been explored, understanding AI in its entirety means that there is still no universal definition (Wirtz et al. 2019). In combination with IoT, AI produces and enhances insights to include home energy management, surveillance, healthcare, and automation (Chen et al. 2019). However, for AI-driven change, decisions regarding products and people require hard trade-offs in present business optimisation strategies (Empowering AI Leadership: AI C-Suite Toolkit 2022).

2.2.2 Big Data (BD)

In a paper by Capei.org 2021, the author highlights how Big Data (BD) supports goal (7) in providing affordable and clean energy via the utilisation of smart meters, thus allowing utility companies to restrict or increase the flow of energy (electricity, gas, and water) and to minimise waste at peak periods in ensuring adequate energy supply (Acquiring Big Data for SDG 7 Through Smart Electric Metering and Grids 2021; Harnessing Big Data for Development and Humanitarian Action 2021; Riekki and Mämmelä 2021). Yapa et al. (2021) describe BD as the management and performing of predictive analytic on large, aggregated datasets using AI and ML algorithms. The value of this data can be of the utmost value to stakeholders and communities, especially in the energy sectors, which include power-producers, financial institutions, investors, consumers, policymakers, and utilities.

In cases such as IoE, grid reliability and stability is effectively maintained through predictive analysis of aggregated data. Sustained energy is then enhanced further by the heterogeneity of data sources, managed and coordinated by measurements accurately obtained via various IoT connected devices (Yapa et al. 2021). However, the collection of aggregate data needs to be protected from information manipulation and malicious attacks that may both leak information to third parties and potentially sabotage the overall grid operation. BD, therefore, plays an important part in the practical management and drive of future grid applications.

2.2.3 Blockchain (BC)

In Kim and Huh, 2020, a BC is referred to as a data tampering prevention technology that stores managed data in a distributed 'block' environment, where small data provides links in chains via P2P, with no arbitrary access to instigate modifications, whilst remaining accessible to anyone. All transactions are recorded in advance of the previous block and distributed equally. As an alternative to keeping records on a centralised server, all BC transactions and records are compared in preventing forgery, then shown to all users (Kim and Huh 2020).

To add new data in a BC, a new block must first be created, announced to the other nodes, then becomes part of every BC instance. However, in this distribution feature, the possibility to link to a competing and/or predecessor block, or to not be appended, means that all nodes must agree on a consensus mechanism to ensure that agreement is reached in regard to the state of a new block, and whether it is permissioned (private) or permissionless (public) (Przytarski et al. 2022).

From cryptocurrencies to general purpose use, BCs have evolved for those that require data integrity and high service availability. In their increased utilisation, BC-based solutions could reinstate the trusted broker, whether in premises or via the cloud (i.e., Data Centre). Therefore, a cloud-based BC mechanism could potentially bring a trusted element to the cloud instead of utilising nodes on premise. For example, the public may utilise multiple entities which then become inherently a trusted source of truth/broker, however, in this case, this also allows vulnerabilities from human error to render the system unreliable. This challenge then comes at a cost in that implementing a BC in substitution to bureaucracy and fallible human, which does not eliminate risk, it simply becomes shifted to other areas (Clavin et al. 2020).

2.3 Application

2.3.1 People: Health Surveillance

The *Smart Future of Healthcare* (SFH) report in 2020 pertains to potential barriers in scaling-up smart meter distribution for health and care monitoring, a major step in meeting SDG (3) Good Health and Wellbeing, for example. Up until recently, smart meters were utilised for consumers in optimising household energy and for utility companies to provide grid reliability, accurate billing whilst reducing operational cost (Theusch et al. 2021) at the same time as meeting decarbonisation objectives. For example, by the end of 2024, the UK aims to have 85% of its consumers equipped with smart meters (Using Smart Meters in Health and Care Monitoring Systems 2021).

The SFH report goes further and states that '*If used as a health and care monitoring technology, the smart meter could soon become a virtually ubiquitous telehealthcare solution. No other Ambient Assisted Living (AAL) or telecare technology comes close for scalability*'. Additionally, four new applications of smart meters in health and care monitoring are identified from the SFH, to include:

- *The mobilisation of social capital, freeing-up healthcare capacity and to improve overall social services.* Smart meter data can identify unhealthy homes and cold dwellings (i.e., similar to weather data), which in turn informs governments in providing efficient homes to their citizens, with the ability to also retrofit Heating Ventilation and Air Conditioning (HVAC) systems to aid better housing conditions. This also means that a government can utilise this data to identify emergency and low credit households and provide support alongside social services in the form of grants (Using Smart Meters in Health and Care Monitoring Systems 2021).
- *Home energy uses data to check on health status* and changes to the general wellbeing of the occupants of a given house (Yassine et al. 2017). This concerns health deterioration and the provision for swift action and response in the case of deviation from typical energy consumption patterns (i.e., the elderly). This approach utilises a remote Non-Intrusive Load Monitoring (NILM) of energy consumption (Theusch et al. 2021), alongside additional appliances monitoring such as a microwave, a toaster, a kettle, an oven, and a washing machine, all of which can indicate/draw inferences through NILM.
- *The use of ML, behaviours, and activities over time* aid computerised systems in identifying unexpected inactivity. For example, sleep disturbances may be inferred by utilising a kettle and additional appliances at night, thereby detecting potential mental health, arthritic pain, and other neurological disorders. Additionally, forgetting to turn off appliances repeatedly, and the increase in late evening energy consumption may indicate restlessness, agitation from confusion, and/or symptoms of dementia (i.e., Sundowning Syndrome) and memory problems. These symptoms can be attributed to mild cognitive impairment, of which the system automatically updates and informs relatives in the case of an elderly person who lives alone (Yassine et al. 2017).
- *Smart Home Energy Management Systems (HEMS)* can provide capabilities regarding self-monitoring (Riekki and Mämmelä 2021) and providing a 'feeling of wellbeing and safety to users' via alerts that include leaving various appliances unsupervised, thus providing safety to homeowners (Smart Future of Healthcare 2021).

2.3.2 Energy: UK Power

In the IoT and BD context, Smart electric meters report information that is produced monthly (also by non-smart meters), reporting every 15 min and making up at total of 2880 reports per household. This huge amount of data made up of data variety, velocity, and volume are typical traits of BD. The first country to use this technology in a commercial setting was started by a Japanese company called Informetics, a subsidiary of Sony and in partnership with Tokyo Electric Power Company. A European version was planned for 2021 following field trials (Smart Future of Healthcare 2021).

With the emergence of the IoE, this places further requirements on traditional grid infrastructure to integrate heterogeneous sources of energy, network distribution management, grid intelligence with BD and AI management (SDG 9). As a means to enhance autonomy and decentralisation of grid operations, disruptive technologies such as BC have been identified to enable self-operated and reliable energy delivery (Yapa et al. 2021). However, for power companies earlier this year, the electricity distribution network operator, UK Power, for example, was trialling and developing Machine Learning (ML: a branch of AI) software and tools to identify, support, and enhance targeted infrastructure investment. This software-based approach provides additional fresh insights, visibility, and flexibility in planning, therefore, effectively being able to respond to market signals and distribution-level conditions (Case Study UK; Nabavi et al. 2021).

2.4 *Summary*

As the USA and UK bring focus and investment upon reducing global GGEs with urgent action efforts against global climate change, policy efforts, national grid resilience and sustainability strategies, enhanced with AI and BD, can help in the fight to mitigate many of the challenges set out by the UN SDGs. For example, with the UK implementing Smart Meters gives hope with regard to meeting decarbonisation objectives, net zero emissions, and all supported via intelligent networks (IoE) with distribution management. However, with significant economic growth, traditional infrastructures require significant investment and innovation to help change human everyday lives in the provision for technology, heat, transport, and communication to be felt on a global scale in meeting SDG 7 (Affordable Clean Energy) and 8 (Decent Work and Economic Growth) objectives. In addition, grid resilience strategies and national security could be said to be open to cyber-attacks due to the compounding shift in technological infrastructure, policy, and leadership, thus presenting challenges in meeting SDGs regarding 7 (Affordable and Clean Energy). The main goal for Cyber-leaders, for example, is to instigate, develop, trust, and create strong partnerships within their respective ecosystems (Global Cybersecurity Outlook 2022) in meeting UN obligations and policies set out by the Paris Agreement 2015 and to include cyber-criminality mitigation strategies.

Utilising SDGs (overall objective: Net Zero by 2050) looks towards a better future; however, any significant gains tend to be undermined by governance and conflict compounded by climate change. As well-established research on energy security is known, in conceptual analytical research regarding sustainability transitions, little attention has been paid to the security threats that influence security policy and transitions as part of a mixed policy approach (Kivimaa and Sivonen 2021; Dong and Zhang 2021).

3 Research Challenges

In a paper by Kivimaa and Sivonen, 2021, of the Finnish Environment Institute SYKE, Climate Change Programme, the authors look at policy coherence and integration analysis of energy and security strategy documents to include sustainability transition research, and how energy niches and landscape pressures are presented via documents concerning Estonia, Finland, and Scotland during 2006–2020. Their findings indicate that energy and security policies show a functional overlap, however, policy coherence and integration are not adequately addressed due to coexisting security considerations.

In conclusion, the authors state that in an increased and more multifaceted landscape produces a more complicated environmental policy (low carbon and hydrocarbon-based), thus coherence becomes more difficult. However, in accelerating such energy transitions, security implications regarding energy niches do not receive sufficient attention (Kivimaa and Sivonen 2021).

3.1 Policy

3.1.1 People: Data Protection

Inclusive of the modern energy markets, consumer data security and privacy present immense challenges alongside AI deployment in smart meters and smart city businesses, primarily due to third parties. As part of a solution in protecting consumer data, the General Data Protection Regulations (GDPR) can be said to worsen the data acquisition methods required for delivering a less intrusive method of data gathering with smart meters (Martins et al. 2022). GDPR requires consent from data subjects to businesses, however, obtaining true consent can be difficult due to a data subject fearing negative consequences (Human-Centred Artificial Intelligence for Human Resources 2021). Therefore, the frontier of data gathering presents challenges between third parties and service providers in whether or not they can access and use the data effectively.

3.1.2 Energy: Predictability

In mitigation of these types of challenges, Torino in Case Study (UK) states that as the UK transitions to a renewables-based energy future, facilitates in its legal obligations and targets for net zero by 2050, major power generation changes mean that effectively managing/responding in a reliable, safe, and cost-effective network is key. Also, due to minimal historical electrical load data, there is limited load visibility (collected data: flow vs time vs demand) that add difficulties.

Due to the increase in low carbon technologies in providing electricity for transport and heat, energy network operators and distribution networks need to acquire more information to keep demand predictable as in the past (Chen et al. 2019). However, because of low carbon demands and intermittent renewable energy sources, this presents challenges in future predictability and reliability of energy distribution. These factors continue to challenge planning network interventions and new connections (new infrastructure) for network operators in the face of high degrees of uncertainty (Case Study UK; Nabavi et al. 2021).

3.2 *Technology*

3.2.1 People: Scalability

Regardless of smart meter benefits for health monitoring and improving care services, potential barriers exist in preventing any real scalability or technology adoption, therefore, causing a lack of collaboration and proposition in presenting progress. As a result, potential lack of funding and public interest may prevent countries like Scotland and Wales from participating in an equitable energy market. If collaboration from within energy, healthcare, and computer science is missing alongside government interest, for example, then studies will only remain theoretical and without evidence of validity and scalability.

3.2.2 Energy: Security

UK Power Networks comprise one of the 14 UK network operators, responsible for overhead and underground power line maintenance and upgrading for: London (*'is one of the most heavily population regions of the UK'* (Annual Population Growth in the United Kingdom in 2020 2021)), South and South East England (UK Power Networks 2021). As a power company ascertains the power grid security state (overall grid condition), non-linear state recovery techniques undertake tasks to assess and predict grid loads, identifying structural abnormalities, updating pricing policies and to inform user controls (Lu et al. 2021; Nabavi et al. 2021). Instances of bad data either from random (i.e., sensor failures), structural (i.e., cyber-attacks [false data injection]) or in successfully identifying the state of a power system continue to present fundamental challenges.

For example, detecting structured and random data is undertaken via state estimators (recovering phase angles and bus voltages) and algorithms in leveraging data collections that use multiple measurement units to include topological and dynamic information. Moreover, identifying bad data and choosing an appropriate protocol for distorted data and performance limits present unknown risks which affects the ability to effectively recover from a system's infrastructure failure or from a potential cyber-attack (Wylde et al. 2020; Tajer et al. 2019; Dong and Zhang 2021).

3.3 Applications

3.3.1 People: Smart Meter Connectivity

Further technological and connectivity issues remain slow in most countries; this presents challenges in terms of smart energy data in health and care monitoring. The UK, even with a fast programme deployment, tended to use smart meters less per household due to connectivity issues, therefore, inadvertently reverting to utilising a more basic mode of operation upon the switching of energy suppliers. Further issues exist such as the installation of Wi-Fi and cellular connectivity in rural areas (i.e., Wales and Scotland), where intermittent or non-existing connectivity prevents accurate and efficient smart meter readings. However, the UK as a whole has improved cellular and Wi-Fi access for elderly people, a main target for new use-cases of smart meter applications (Smart Future of Healthcare 2021).

3.3.2 Energy: Market Supplier

One of the main challenges here in regard to UK Power Networks being one of the 14 networks in the UK is that per energy supplier, and their different AI strategies, modernisation is needed (Dong and Zhang 2021). For example, the Welsh energy company Wales-West-Utilities' strategy does not take into consideration the overlap between the more modern UK suppliers' policy innovations. This may contribute weaknesses in data infrastructure and data sharing, therefore, present unfair competition, inaccurate data gathering practices (Smart Meter), and put Wales at a competitive disadvantage (Digitalisation Strategy: A Digital Vision for Safe and Resilient Networks 2021).

3.4 Summary

As shown above, due to energy and policy overlap that present security challenges when concerning data security and privacy, coupled with limited access, all contribute to problems in transitioning to renewable-based energy practices. Also, due to minimal historical load data, visibility of the grid state is limited which present additional challenges in predictability, especially with the intermittent nature of transition to renewable energy sources. In addition, connectivity issues further compound modernisation (i.e., smart meters) and technology adoption that inevitably causes data infrastructure weakness in data sharing. With the UK enjoying multiple energy companies, all with separate AI strategies, and all simultaneously being activated at different times, this may cause significant service disruption, health inequalities and highlight additional and compounded discrimination practices in terms of smart meter equality and energy equity (SDG 7). In addition, due to geographical and topological factors, cellular and Wi-Fi connectivity remain a constant challenge for rural communities. In terms of data protection, BC may provide effective resource accountability and transparency, however, without abundant digital infrastructure, these technologies remain out of reach for communities that are located away from cities and suburbs.

4 Potential Solutions

4.1 Policy

People: Data Protection Taking into account all the above-mentioned challenges, utilising BC provides a method in ensuring that availability of data will always be verifiable and trustworthy (Clavin et al. 2020), especially with a robust and prescribed agreement such as a Smart Contracts (SC). However, in Ng et al. (2021), challenges include important aspects of privacy in digital solutions: anonymisation and healthcare data techniques, ethical issues with the information used in fighting the pandemic, public safety and privacy with contact tracing applications (Majeed and Hwang 2021). In addition, there are not enough studies that are BC-based to have concrete evidence as an alternative, thus no superior results as BC technologies mature in contrast to traditional methods (Aysan et al. 2021).

Energy: Climate Change—Blockchain: Interoperability To address the challenges of climate change, distributed renewable energy sources that balance production, transmission, and distribution are in high demand (Aklilu and Ding 2022). Over the last decade with increases in global energy demand and sustainable development, these challenges can be mitigated through IoE, BC technology with SCs and implemented further by micro- and nanogrids in hybrid power systems.

4.2 Technology

In Yassine et al. (2017), the authors highlight how the advancement of BD mining technologies (processing large data sets for actionable insights) can be used in understanding how people go about their business in their daily lives. As one of the most challenging factors is affected by influxes of people to city centres, hence major investment in digital transformation is continually underway in sustaining healthy ecosystems for city populations. A key factor in maintaining accurate information of local government dissemination of resources is Electronic Health Records (EHR), of which BC interest for biomedical applications has doubled every year since 2015–2019. Although the potential of BC is high, five characteristics must be factored in for any BC solution (Clavin et al. 2020).

- Security
- Scalability
- Privacy
- Interoperability
- Governance

To achieve these BC characteristics and needs, therefore, require:

- Nonrepudiation
- Auditability
- Transparency
- Decentralisation
- Distribution
- Cryptography
- Immutability

Blockchain: Decentralisation Clearly, carbon markets are being developed globally as a consequence of the 2015 Paris Agreement, additionally BCs (also named as public exchange book) do not require a central manager, it is managed by all traders/countries separately, as in the past, banks owned all trading books, all transactions were made via banks which caused issues in the event of hacking. As BC has no central management system, it is also safe from hacking as all traders independently have constantly updated books of their own (Kim and Huh 2020).

Blockchain: Auditability Alladi et al. (2019), highlight how the trading and consumption of electricity give rise to serious privacy and security challenges in the context of the smart grid. Alongside the integration of the Internet of Things (IoT) and wireless sensor networks, these challenges [through the smart grid] are set to aid in future electric supply.

Blockchain: Immutability For example, BC if utilised in a credit-based payment scheme via peer-to-peer energy trading can greatly enhance the process. Data aggregation BC systems that can also effectively resolve security and privacy issues in the grid. Additionally, energy companies and distribution systems may utilise BC by remote control for energy flow in a given area whilst being informed by the relevant area statistics and to target smart grid maintenance and diagnosis tasks (Alladi et al. 2019). Kim and Huh (2020), also bring attention to how carbon credits can help mitigate environmental pollution [i.e., carbon emissions] as it becomes a more serious issue in 2021 on-wards, therefore, carbon trading systems will be applied and available for individuals.

Blockchain: Governance For this to be realised, however, the exchange of credits is needed. The authors propose a more reliable BC mechanism (dApp) to measure carbon emission rights via the UN SDGs'17 goals/tasks. The BC-based carbon emissions rights verification mechanism utilised governance system analysis and BC mainnet engine alongside BD and AI in mobile cloud environments (Kim and Huh 2020).

Blockchain: Transparency With regard to awarding government contracts in smart meter roll-out schemes (including Covid-19 mitigation procurement), BC could potentially help bring into focus corruption risk factors by providing evidence of tamper-free (oversight) transactional data for increased uniformity and objectivity via SCs. This enhances transparency between state and non-state actors with accountability at the fore.

Blockchain: Security There is no simple answer here, BC technology would seem more effective, especially for SDG (9) Industry Innovation and Infrastructure, if a more holistic interface system is generated from the outset, therefore, becoming a more robust overall strategy in the effective distribution of procurement for government spending in times of crisis.

Blockchain: Interoperability For governments that can utilise smart meter data to identify emergency and low credit households and provide support alongside social services in the form of grants, the process is often inefficient, convoluted, and opaque, causing money to be lost from middlemen, banking fees, and possible corrupt financial diversions. BC may be utilised to build public trust (Olivares-Rojas et al. 2021) and to dis-intermediate and minimise actor numbers that are involved and, therefore, minimise cost (Olivares-Rojas et al. 2021), opportunities to illicit financial siphoning and to streamline the overall process.

However, for users and recipients to effectively manage a BC-based system, this may prove challenging for less resourced and technologically savvy organisations and individuals and cause discrimination with exclusion from the system if unable to utilise the technology. This flies in the face of SDG (10) Reduced Inequalities. Although not a holistic solution, BC-based disbursement systems can help identify and mitigate corrupt practices that frequently arise in the human aid context (Blockchain Alone Can't Prevent Crime 2021; Olivares-Rojas et al. 2021).

Blockchain: Data Privacy and Security According to the UK Public Interest Advisory Group (PIAG), smart meter data access for public-interest purposes indicates that consumers struggle to identify the benefits and risks of their data, therefore, find smart meter data a less sensitive method of sharing data with companies. This, in turn, would aid third parties in providing services such as health and care monitoring (Using Smart Meters in Health and Care Monitoring Systems 2021). Unfortunately, as mentioned from all the above, the potential for corrupt financial and regulatory oversight practices remains prevalent; therefore, this includes the possibility of data protection breaches, especially with medical health data.

Smart Contracts SCs may benefit consumers here, however, in utilising smart meters alongside BC can help prevent tampering and illicit practices. However, if applications of AI only seek to replace interactive (i.e., voice or alarm) system responses for customer calls, the automation of basic computer tasks may not be as transformational in contrast to applications that can learn and improve performance over time (Wirtz et al. 2019).

4.3 Application

In the context of emission and energy trading systems, the EU utilises a common emissions system, China announced an emission trading system for their power generation sector, Korea has also contemplated using a similar model with Japan and China. Additionally Japan, alongside 17 other developing countries to include Vietnam, has a common offset system; California and Quebec also utilise linked systems (Kim and Huh 2020).

4.4 Summary

The crises of the UK energy and healthcare sectors highlight how significant challenges can compound during global shifts in technologies, political will, and geo-politics and their effects upon the global economy. In trying to initiate change, governments inherently fall short of undertaking these huge tasks due to the short-term nature of governance systems in aiding fresh and up-to-date oversight.

For a more holistic and future-proof model to be successful regarding healthcare, much longer policy implementation, similar to the UN SDGs, is enacted locally to balance data protection and human rights, for example; however, energy infrastructure can provide less intrusive methods of data acquisition, whilst enabling care services and providers to deploy resources more effectively. As with these challenges, the UK energy sector competes individually and initiate separate policies and practices that may present further data protection, health inequalities, and overall consumer equity challenges nationwide.

Customer connectivity and gaining current information present additional issues in terms of aligning BD databases cohesively, and with systems that ensure transparency and accountability, thus helping to build public trust. BC technologies can also provide an alternative approach to these challenges; however, there has to be an incremental application throughout the energy sectors.

This is especially important as this helps provide emphasis in undertaking more partnerships and research across the higher-education and industry sectors in ultimately finding more cohesive strategies in homogenising the national grid (IoE), therefore, truly engaging and partaking in fighting climate change whilst achieving SDGs to enable a more global perspective in effective healthcare data acquisition practices and resource deployment strategies.

5 Conclusion and Recommendations

5.1 Conclusion

As the UK faces rises in energy prices due to political, social, and economic uncertainty exacerbated by the pandemic, global governance and future joint-efforts and strategies tend to be long term. With these efforts in the face of current and real-world economic and health crises, any actions may become negligible. For example, governments and energy companies individually instigate and implement AI strategies, which may provide excellent and effective measures in providing healthcare services from smart data for developed countries. This is only useful for those who have reliable internet access.

However, in developing countries, resources are scarce and in contrast to developing countries, the resources made available from AI investment and Covid-19 mitigation strategies could be allocated more equitably, regardless of ensuring AI leadership and dominance economically. Surveys indicate that consumers and the public are unaware of their data output value and find smart meters less intrusive; however, the data protection ramifications compound and conflict due to individual actors and policy deployment that show little interest in an overall cohesive strategy in terms of equitable health distribution and national security.

In times of crisis in the energy sectors and with the current pandemic, is it really appropriate to take on huge nationwide AI strategies in health data acquisition, whilst simultaneously trying to achieve climate change goals and objectives? From one perspective, the lack of regional, national, and international policy and implementation cohesiveness may provide opportunities for illicit financial and political practices amongst the transitional issues.

Here, the benefits of AI in 'remotely managing chronic diseases, reducing pandemics, and improving food security and sustainable agriculture, to increasing public safety through monitoring infrastructure and providing services to take care of elderly populations' (Wirtz et al. 2019), BC and SCs can also provide a method of transparency and accountability to try an ascertain and grapple with the additional confusion and noise. However, the inherent properties of BC technologies require that implementation is on a much more grand scale, and in step with human activities concerning the public and private appetites for accountability being a force for equality and justice.

5.2 *Recommendations*

As the UK, amongst other nations experience vast uncertainty in the global economies of scale context, the political will to remain a viable candidate on the world stage compound the duties of governance at the country and/or societal level. Leadership requires that successful governance can come from prioritisation of resources, and to effectively lead is better than contributing more to the political noise.

Additionally, the funding of AI strategies vs aiding developing countries should be balanced in a more equitable way, thereby facilitating a longer-term strategy in view of new and potentially larger customer bases globally. Trying to assume more AI technology dominance from an already dominant position, in the short term may seem popular vs a given country and their voters, however, in utilising AI to identify global humanitarian aid hot-spots could be more beneficial in the long term economically.

Furthermore, the protection of civil liberties and data sharing should be made more of paramount importance thorough education. Due to the numerous global crises, this has taken a back seat and could be seen as a data 'free-for-all' for large actors and non-state actors. The lack of education in developed countries regarding data is surprising and could allow further economic and social sanctions long term, inadvertently driven by this phenomenon. From a small to large scale point of view, BC can help mitigate data and cyber-threats, but only if the population it serves can appreciate the implicit and inherent value of data and its acquisition, use, and its resource deployment strategies (i.e., health outcomes).

References

Acquiring Big Data for SDG 7 Through Smart Electric Metering and Grids. https://cepei.org/en/documents/big-data-smart-metering-grids/. Accessed 30 Dec 2021.

Aklilu, Y. T., & Ding, J. (2022). Survey on blockchain for smart grid management, control, and operation. *Energies, 15*(1), 193.

Alladi, T., Chamola, V., Rodrigues, J. J. P. C., & Kozlov, S.A. (2019). Blockchain in smart grids: A review on different use cases. *Sensors, 19*(22), 4862.

Annual Population Growth in the United Kingdom in 2020, by Region. https://www.statista.com/statistics/294681/uk-population-growth-by-region/. Accessed 30 Dec 2021.

Aysan, A. F., Bergigui, F., & Disli, M. (2021). Using blockchain-enabled solutions as SDG accelerators in the international development space. *Sustainability, 13*(7), 4025.

Blockchain Alone Can't Prevent Crime, But These 5 Use Cases Can Help Tackle Government Corruption. (2021). https://www.weforum.org/agenda/2020/07/5-ways-blockchain-could-help-tackle-government-corruption/l. Accessed 30 Dec 2021.

Case Study (UK). (2021). Digitalising Energy Systems for Net Zero. https://www.ofgem.gov.uk/publications/case-study-uk-digitalising-energy-systems-net-zero. Accessed 28 Dec 2021.

Chen, Y. Y., Lin, Y. H., Kung, C. C., Chung, M. H., & Yen, I. (2019). Design and implementation of cloud analytics-assisted smart power meters considering advanced artificial intelligence as edge analytics in demand-side management for smart homes. *Sensors, 19*(9), 2047.

Clavin, J., Duan, S., Zhang, H., Janeja, V. P., Joshi, K. P., Yesha, Y., Erickson, L. C., & Li, J. D. (2020). Blockchains for government: use cases and challenges. *Digital Government: Research and Practice, 1*(3).

Climate Action: Tackling Climate Change. (2021). https://www.un.org/sustainabledevelopment/climate-action/. Accessed 30 Dec 2021.

Department of Energy's Electricity Advisory Committee Establishes the Grid Resilience for National Security Subcommittee. (2021). https://www.energy.gov/oe/articles/department-energy-s-electricity-advisory-committee-establishes-grid-resilience-national. Accessed 30 Dec 2021.

Digitalisation Strategy: A Digital Vision for Safe and Resilient Networks. (2021). https://www.wwutilities.co.uk/media/3571/wales-west-utilities-digitalisation-strategy.pdf. Accessed 30 Dec 2021.

Dong, Z. Y., & Zhang, Y. (2021). Interdisciplinary vision of the digitalized future energy systems. *IEEE Open Access Journal of Power and Energy, 8*, 557–569.

Electricity Transmission: Our resilient Whole System approach. (2021). https://www.nationalgrid.com/uk/electricity-transmission/document/136231/download. Accessed 30 Dec 2021.

Empowering AI Leadership: AI C-Suite Toolkit: January 2022. (2022). https://www3.weforum.org/docs/WEF_Empowering_AI_Leadership_2022.pdf. Accessed 26 Jan 2022.

Energy Bills Could Rise by 50% Amid 'National Crisis' of Soaring UK Prices. (2021). https://www.theguardian.com/business/2021/dec/23/energy-bills-could-rise-by-50-amid-national-crisis-of-soaring-uk-prices. Accessed 23 Dec 2021.

Energy Costs and Covid Pose 'existential threat' to UK's Small Businesses. (2021). https://www.theguardian.com/business/2021/dec/30/energy-costs-and-covid-pose-existential-threat-to-uks-small-businesses. Accessed 31 Dec 2021.

Global Cybersecurity Outlook 2022: Insight Report 2022. (2022). https://www3.weforum.org/docs/WEF_Global_Cybersecurity_Outlook_2022.pdf. Accessed 26 Jan 2022.

Harnessing Big Data for Development and Humanitarian Action. (2021). https://www.unglobalpulse.org/wp-content/uploads/2018/05/UNGP_Annual2017_final_web.pdf. Accessed 30 Dec 2021.

HM Government National AI Strategy. (2021). https://assets.publishing.service.gov.uk/government/uploads/system/uploads/attachment_data/file/1020402/National_AI_Strategy_-_PDF_version.pdf. Accessed 10 Dec 2021.

Human-Centred Artificial Intelligence for Human Resources: A Toolkit for Human Resources Professionals Toolkit December 2021. (2021). https://www3.weforum.org/docs/WEF_Human_Centred_Artificial_Intelligence_for_Human_Resources_2021.pdf. Accessed 26 Jan 2022.

Kim, S. K., & Huh, J. H. (2020). Blockchain of carbon trading for UN sustainable development goals. *Sustainability, 12*(10), 4021.

Kivimaa, P., & Sivonen, M. H. (2021). Interplay between low-carbon energy transitions and national security: an analysis of policy integration and coherence in Estonia, Finland and Scotland. *Energy Research & Social Science, 75*, 102024.

Lu, T., Chen, X., McElroy, M. B., Nielsen, C. P., Wu, Q., & Ai, Q. (2021). A reinforcement learning-based decision system for electricity pricing plan selection by smart grid end users. *IEEE Transactions on Smart Grid, 12*(3), 2176–2187.

Majeed, A., & Hwang, S. O. (2021). A comprehensive analysis of privacy protection techniques developed for COVID-19 pandemic. *IEEE Access, 9*, 164159–164187. https://doi.org/10.1109/ACCESS.2021.3130610

Martins, J., Strasser, T. I., & Sănduleac, M. (2022). Guest editorial: Smart meters in the smart grid of the future. *IEEE Transactions on Industrial Informatics, 18*(1), 653–655.

Nabavi, S. A., Motlagh, N. H., Zaidan, M. A., Aslani, A., & Zakeri, B. (2021). Deep learning in energy modeling: application in smart buildings with distributed energy generation. *IEEE Access, 9*, 125439–125461.

Ng, W. Y., Tan, T. E. P., Movva, V. H., Fang, A. H. S., Yeo, K. K., Ho, D., San Foo, F. S., Xiao, Z., Sun, K., Wong, T. Y., Sia, A. T., & Ting, D. S. W. (2021). Blockchain applications in health care for covid-19 and beyond: A systematic review. *The Lancet Digital Health, 3*(12), e819–e829.

Olivares-Rojas, J. C., Reyes-Archundia, E., Gutiérrez-Gnecchi, J. A., Molina-Moreno, I., Cerda-Jacobo, J., & Méndez-Patiño, A. (2021). A transactive energy model for smart metering systems using blockchain. *CSEE Journal of Power and Energy Systems, 7*(5), 943–953.

Positive AI Economic Futures Insight Report November 2021. (2021) https://www3.weforum.org/docs/WEF_Positive_AI_Economic_Futures_2021.pdf. Accessed 26 Jan 2022.

Przytarski, D., Stach, C., Gritti, C., & Mitschang, B. (2022). Query processing in blockchain systems: Current state and future challenges. *Future Internet, 14*(1).

Riekki, J., & Mämmelä, A. (2021). Research and education towards smart and sustainable world. *IEEE Access, 9*, 53156–53177.

Smart Future of Healthcare. (2021). https://2020health.org/wp-content/uploads/2020/11/2020health-SFoH-Full-report.pdf. Accessed 30 Dec 2021.

Tajer, A., Sihag, S., & Alnajjar, K. (2019) Non-linear state recovery in power system under bad data and cyber-attacks. *Journal of Modern Power Systems and Clean Energy, 7*(5), 1071–1080.

The Guardian View on an Energy Price Shock: A Crisis in the Making. (2021). https://www.theguardian.com/commentisfree/2021/sep/20/the-guardian-view-on-an-energy-price-shock-a-crisis-in-the-making. Accessed 21 Dec 2021.

The Paris Agreement. (2021). https://unfccc.int/process-and-meetings/the-paris-agreement/the-paris-agreement. Accessed 21 Dec 2021.

Theusch, F., Klein, P., Bergmann, R., Wilke, W., Bock, W., & Weber, A. (2021). Fault detection and condition monitoring in district heating using smart meter data. In *PHM Society European Conference* (Vol. 6, pp. 11–11).

UK Power Networks. (2021). https://powercompare.co.uk/uk-power-networks/. Accessed 30 Dec 2021.

Using Smart Meters in Health and Care Monitoring Systems. (2021). https://www.smart-energy.com/industry-sectors/smart-meters/using-smart-meters-in-health-and-care-monitoring-systems/. Accessed 30 Dec 2021.

Wirtz, B. W., Weyerer, J. C., & Geyer, C. (2019). Artificial intelligence and the public sector—applications and challenges. *International Journal of Public Administration, 42*(7), 596–615.

What Caused the UK's Energy Crisis? (2021). https://www.theguardian.com/business/2021/sep/21/what-caused-the-uks-energy-crisis. Accessed 21 Dec 2021.

Wylde, V., Prakash, E., Hewage, C., & Platts, J. (2020). *Data cleaning: challenges and novel solutions*. Cardiff: Cardiff Metropolitan University. AMI 2020 Presentation.

Wylde, V., Rawindaran, N., Lawrence, J., Balasubramanian, R., Prakash, E., Jayal, A., Khan, I., Hewage, C., & Platts, J. (2022). Cybersecurity, data privacy and blockchain: A review. *SN Computer Science, 3*(2), 1–12.

Yapa, C., de Alwis, C., & Liyanage, M. (2021). Can blockchain strengthen the energy internet? *Network, 1*(2), 95–115.

Yassine, A., Singh, S., & Alamri, A. (2017). Mining human activity patterns from smart home big data for health care applications. *IEEE Access, 5*, 13131–13141.

The Use of Artificial Intelligence in Content Moderation in Countering Violent Extremism on Social Media Platforms

Kate Gunton

Abstract This chapter critically evaluates the use of artificial intelligence (AI) in content moderation to counter violent extremism online. To this end, this chapter focuses on measuring the accuracy of AI in content moderation, the occurrences of false positives and false negatives and the infringements on the freedom of expression and democracy. This chapter also presents a critical analysis into the use of de-platforming measures in content moderation. A critical discussion is provided on far-right violent extremists migration from mainstream social media platforms to alt-tech platforms and the use of AI in the de-platforming process. It is argued that the use of automated content removal is limited in effectiveness, and that the use of AI in content moderation could lead to the violation of the principles of freedom of expression and democracy. Furthermore, this chapter highlights the value of de-platforming measures as a more effective tool to counter violent extremism online. It also considers the benefits of the use of AI in the de-platforming measures to enhance the detection of violent extremist users and networks online.

Keywords Artificial intelligence · Content moderation · Countering violent extremism · Social media platforms · Human rights · Freedom of expression · Content removal

1 Introduction

The continuous evolution of the internet has created substantial challenges for governments around the world when striving to counter violent extremism (CVE) (Piazza & Guler, 2019). Violent extremist groups tend to exploit the internet, particularly social media platforms, by rapidly disseminating extremist narratives and propaganda to large-scale audiences with the aim of recruiting and radicalising violent extremist actors, funding their harmful ideologies, and planning potential

K. Gunton (✉)
Independent Researcher, Wales, UK

R. Montasari (ed.), *Artificial Intelligence and National Security*,
https://doi.org/10.1007/978-3-031-06709-9_4

attacks (United Nations Office on Drugs and Crime, 2012). The growing presence of violent extremist groups on mainstream social media platforms, such as Facebook and Twitter, has put pressure on many governments to re-examine their CVE online responses – with many choosing to adopt tougher content removal measures (Guhl et al., 2020). For example, the German government enacted the *Netzwerkdurchsetzungsgesetz* (*NetzDG*) law in 2017, permitting fines of up to 50 million euros to be imposed on social media platforms with over 2 million users for failing to remove illegal content, such as extremist images and propaganda (Miller, 2017).

Additionally, the UK government published the Online Harms White Paper in 2019 which highlighted the need to introduce an alternative regulatory framework to address online extremism – one that improves the safety of online users by placing a statutory duty of care on social media companies (HM Government, 2019). In response, in 2021, the UK government drafted an Online Safety Bill, permitting fines of up to 18 million pounds on social media companies or 10% of their annual turnover for failing to remove harmful content, including extremist images and propaganda, for user's protection (Wakefield, 2021). With some governments adopting stricter regulatory frameworks on content removal, social media companies are required to detect and remove the content in a very small timescale (Gorwa et al., 2020). Social media companies are turning to artificial intelligence (AI) processes to aid in content moderation to detect and remove the vast amounts of harmful violent extremist content being uploaded and disseminated across their platforms each day (Llansó et al., 2020). AI in content moderation can refer to a range of automated processes (Llansó et al., 2020) – from predictive machine learning (ML) tools to automated hash-matching (Gorwa et al., 2020).

This chapter seeks to critically evaluate the use of AI in content moderation to counter violent extremism online, focusing on measuring the accuracy of AI in content moderation, the occurrences of false positives and false negatives and the infringements on the freedom of expression and democracy. There will also be a critical evaluation on the use of de-platforming measures in content moderation to counter violent extremism online, focusing on far-right violent extremists' migration from mainstream social media platforms to alt-tech platforms and the use of AI in the de-platforming process. Ultimately, this chapter will argue that the use of automated content removal to counter violent extremism online is limited in effectiveness as it does not take contextual information into account, causing inaccurate applications, such as false positives, and cannot accurately flag certain extremist content. It also argues that the use of AI in content moderation violates the principles of freedom of expression and democracy, which is intensified by the lack of transparency and accountability for social media companies. Finally, this chapter argues that de-platforming measures are a more effective tool to counter violent extremism online and that it is worth considering the use of AI in the de-platforming measures to enhance the detection of violent extremist users and networks online.

2 Definitions

2.1 Violent Extremism

There is a lack of consensus around the definition of violent extremism among policymakers and academics. Extremism has been defined by the UK Government as the 'vocal or active opposition to fundamental British values, including democracy, the rule of law, individual liberty and mutual respect and tolerance of different faiths and beliefs' (Home Office, 2015). However, this definition of extremism has received a significant amount of criticism. One of the main criticisms is that the UK Government is considerably vague and ambiguous when defining what constitutes fundamental British values (Lowe, 2017), which could lead to the stigmatisation of certain communities (Vincent & Hunter-Henin, 2018). It has also been criticised for failing to take into account far-right extremist groups who are intolerant of different races and ethnicities and, as a result, enables them to gravitate further into nationalism (Allen, 2021). According to Kundnani and Hayes (2018), there is a general lack of consistency when defining extremism, which can have significant impacts on the creation and implementation of CVE policies.

There are conceptual differences between definitions of extremism that are behavioural – which focus on the measures taken by actors to reach a political objective – and idealistic – which focus on active opposition to a society's core values (Stephens et al., 2018). Neumann (2013) points out that most definitions of 'non-violent extremism' take an idealistic perspective focusing on the extreme ideology itself, compared to definitions of 'violent extremism' that take a behavioural perspective focusing on the use of violent methods to achieve a political goal. However, there have been criticisms that differentiating between violent and non-violent extremism is ineffective because it is possible that an individual can hold extremist views without participating in violent measures to achieve a political goal, only to engage in violent acts when an opportunity arises (Schmid, 2014). Nevertheless, there will be a differentiation between violent and non-violent extremism within this chapter as the critical evaluation is focused specifically on the use of AI in content moderation to counter violent extremism online. Thus, violent extremism will be defined as 'encouraging, condoning, justifying, or supporting the commission of a violent act to achieve political, ideological, religious, social, or economic goals' (Federal Bureau of Investigation (FBI), n.d., para. 1).

2.2 CVE

CVE will be defined as a range of soft, non-coercive responses taken to combat the underlying causes of violent extremism, for example, radicalisation (Selim, 2016). CVE responses are mostly centred around the concept that 'rehabilitation, reintegration, or prevention' will reduce violent extremism (LaFree & Freilich, 2019, p. 13.2).

3 AI in Content Moderation

3.1 *Measuring the Accuracy of AI in Content Moderation*

Automated predictive ML tools are trained to identify and differentiate between a variety of content based on specific datasets of information (Llansó et al., 2020). In 2018, the UK government announced that a brand-new ML tool had been developed by ASI Data Science and the Home Office, which was able to detect and analyse the sounds and imagery in Islamic State (IS) propaganda videos (Home Office, 2018). The research found that the new ML tool could automatically detect 94% of IS content – with over 99.9% accuracy (Home Office, 2018). Similarly, between June and December 2017, YouTube reported similar levels of success with 98% of violent extremist content automatically detected with almost half removed within 2 hours of being posted (Wojcicki, 2017). Despite appearing incredibly effective on the surface level, reports on the success of automated content moderation are arguably misleading (Gillespie, 2020). The bulk of content being detected and removed are duplicates of content that have previously been assessed by human moderators, conveying that they are not accurately detecting emerging harmful content (Gillespie, 2020). It has also been argued that specialised ML tools created to detect and remove duplicates do not identify the same content in multiple contexts, such as violent extremist propaganda being reported in the news (Llansó, 2020), as they are limited to specific datasets of information (Duarte et al., 2018). Likewise, hash-matching and keyword filters have been criticised for not considering different contexts, and some argue that they are susceptible to making overbroad applications when seeking out duplicates (Engstrom & Feamster, 2017). Thus, it can be argued that using automated content removal measures to counter violent extremism online is limited in effectiveness as the lack of contextual awareness can produce inaccurate and overbroad applications.

When measuring the accuracy of AI in content moderation, it is important to consider the occurrences of both false positives and false negatives and how they are dealt with by social media companies (POST, 2020). A false positive refers to the automatic removal of genuine content from a social media platform and a false negative refers to a system identifying harmful content as genuine (Ofcom, 2019). It is impossible to concurrently decrease the likelihood of both false positive and false negative occurrences in content moderation and, as a result, social media companies are faced with having to prioritise the least impactful outcome (United Nations Office of Counter-Terrorism (UNCCT) & United Nations Interregional Crime and Justice (UNICRI), 2021). Although both outcomes have significant consequences, increasing the likelihood of false positives minimises the risk of harmful violent extremist content escaping detection (UNCCT & UNICRI, 2021). However, there are a variety of different guidelines and standards that social media companies must abide by when removing extremist content from their platforms that may impact the removal of innocent content, as well the difficulty of abiding by the differing definitions of extremism depending on the law of the state in which they are situated (van der Vegt et al., 2019). False positives and false negatives are inevitable when there have been failures to develop a clear and exact definition of the content needing to be targeted (Duarte et al., 2018). It can be suggested that filters using imprecise keywords or terms, such as 'support' or 'glorification', should be avoided as they are vulnerable to misinterpreting certain contexts, for example, where individuals are expressing sympathy (Díaz & Hecht-Felella, 2021). Ofcom (2019) argues that it is essential for content moderation to be contextually aware, such as cultural and legal differences, when operating on a global scale to be effective which would, thus, reduce the number of false positives. Yet, it is arguably impossible to address the issue of false positives and false negatives when there is a lack of consensus between governments, law enforcement agencies and social media companies on what even constitutes 'extremist' content (van der Vegt et al., 2019).

There are additional concerns that most far-right extremist content is not directly associated with one group, which creates difficulties from the outset when detecting content for removal as it may not be flagged instantly (Conway, 2020). With the rise of far-right rhetoric in many Western countries, particularly amongst some political leaders, it has become increasingly challenging to differentiate between far-right and far-right violent extremism in content (Ganesh & Bright, 2019). The use of memes is prevalent within online far-right extremist communities due to their indirect meanings and interpretation which, ultimately, prevents automated systems from flagging it as extremist content (Lee, 2020). Due to the difficulties flagging memes, there is more time for violent extremist users to disseminate the meme across the platform and, consequently, allows the ideology reflected in the content to become normalised amongst other users (Liang & Cross, 2020). According to Weimann and Am (2020), ambiguous visual cues are almost impossible for AI systems to detect and even pose challenges for human moderators as it requires comprehensive knowledge of the violent extremist group's ideology messaging tactics. Thus, the use of AI in content moderation is limited in effectiveness as it cannot accurately flag certain extremist content.

3.2 *Infringements on the Freedom of Expression and Democracy*

The use of AI in content moderation has raised concerns about infringements on the fundamental principles of freedom of expression and democracy (Llansó et al., 2020). For example, Germany's NetzDG law has especially encountered significant criticisms for its violation of the freedom of expression (Tworek & Leerssen, 2019) as it 'poses a threat to open or democratic discourse' and encourages the removal of legal content which limits the users right of free speech (Global Network Initiative (GNI), 2017). Additionally, other critics have expressed concerns that the NetzDG law may result in social media becoming privatised to law enforcement when assessing the legality of content (Federal Cabinet, 2017). However, Heiko Maas (as cited in Tworek & Leerssen, 2019) argued that Germany's Nazi history and use of militant democracy has demonstrated that free speech can be restricted to protect the norms of a democratic state. Although being able to justify automated content removal without infringing on freedom of expression becomes increasingly difficult (Ganesh & Bright, 2019), it may be necessary in some circumstances to protect and safeguard other online users from harm. Despite this, the NetzDG law, along with other strict regulatory frameworks on content moderation, arguably undermines the basic principles of democracy as social media companies are left to makes decisions on what is considered extremist content which can often significantly differ from that of a democratic state (West, 2021). To resolve this issue, it has been suggested that the development of new regulatory structures, such as 'e-courts', to oversee decisions on how certain types of content should be moderated, which would reduce the power of social media platforms over their user's expression and, ultimately, promote democracy (Centre for Data Ethics and Innovation, 2020).

Concerns surrounding automatic content removal's infringements on the freedom of expression are amplified by the lack of transparency and accountability for private companies during the decision-making process of content removal (West, 2021). The Centre for Data Ethics and Innovation (2021) suggested that social media platforms need be more transparent by releasing information on the accuracy of their automatic content removal process, and how they decide what is considered prohibited content. It should be noted that certain types of ML tools do not record the processes involved when decisions on removing potential violent extremist content are made (Coeckelbergh, 2019). As a result, in some cases in which an individual seeks an appeal for a false positive removal, the reason for the takedown will not be evident as the automated decision-making process is indecipherable (Henschke & Reed, 2021). This is incredibly problematic as the decision-making process behind the automated content removal must be scrutable by both the social media companies and those individuals who have appealed (Henschke & Reed, 2021). Thus, using automated content removal measures on mainstream social media platforms to counter far-right extremism online is ineffective as it undermines both the basic principles of freedom of expression and democracy.

4 The Use of De-platforming Measures

De-platforming refers to the temporary block or permanent removal of users, or dangerous organisations and groups, from social media platforms due to violating the companies' guidelines and standards (Rogers, 2020). The use of de-platforming has been criticised for encouraging violent extremists, particularly far-right violent extremists, to migrate from mainstream social media platforms to alt-tech social media platforms as it provides a protected space for far-right extremist ideology to be distributed (Donovan et al., 2018). Violent extremist groups do not need to abide by strict regulatory frameworks on content removal when they inhabit smaller, alt-tech platforms (Guhl et al., 2020). Nouri et al. (2021) argue that de-platforming far-right extremist groups prevent mainstream social media platforms from being used as a way of directing their members to less moderated platforms and limits the number of potential recruits. In March 2018, far-right extremist group Britain First was removed from Facebook for violating 'Community Standards' by engaging in hate crime, but it migrated over to the alt-tech platform Gab (Nouri et al., 2019). Despite migrating to a smaller platform with fewer users, Nouri et al. (2019) found that there was increased engagement with extreme content by Britain First members due to the lack of censorship. This indicates that the expression of extremist views had previously been restricted by harsher moderation on mainstream social media platforms. Additionally, the findings of Nouri et al. (2021) demonstrate that the absence of regulation on the alt-tech platform Gab enables a shift to the normalisation of hateful extremist content, which highlights the ineffectiveness of the selective nature of the NetzDG law and other legislative attempts to regulate extremist content mainstream social media companies. However, Rogers (2020) found that, despite far-right extremist celebrity activity remaining steady on the alt-tech platform Telegram, there was a decrease in the size of the audience and the language employed became less extreme and harmful. Similarly, Guhl et al. (2020) found that a sample of 25 far-right extremist groups had just over 10% of followers from mainstream social media platforms who had migrated to alt-tech platforms, indicating that de-platforming far-right groups reduces their reach and does not directly displace users to alternative platforms. Furthermore, Guhl et al. (2020) found that groups with a presence on mainstream social media platforms had higher numbers of followers on alt-tech platforms than groups that had been banned from the mainstream. Thus, this suggests that de-platforming far-right extremist groups from mainstream social media platforms displaces a small percentage of users to alternative platforms, but effectively reduces the overall reach of far-right extremist narratives. Despite this, Guhl et al. (2020) emphasise that more empirical research is needed to determine the extent of the impact that de-platforming measures have had on reducing the exposure of mainstream social media users to online violent extremists.

Due to the effectiveness of de-platforming measures in CVE, it is important to consider the use of AI in the de-platforming measures as it may further enhance the process. ML tools are frequently used to detect and remove violent extremist content and propaganda from mainstream social media platforms, but there is a need to incorporate these automated systems into de-platforming processes (Chaabene et al., 2021). Although there are existing approaches enabling the automatic detection of users displaying extremist behaviours, there are limited techniques to analyse the network of the user to identify groups (Chaabene et al., 2021). Chaabene et al. (2021) proposed an ML model which would provide the necessary tools to detect and predict violent extremist behaviour based on their content and to identify extremist networks using a Twitter's graph.

5 Conclusion

Having critically evaluated the use of AI in content moderation, it can be concluded that utilising automated content removal to counter violent extremism online is limited in effectiveness as it does not take contextual information into account, causing inaccurate applications, such as false positives. Social media platforms must be aware of the importance of cultural and legal differences between the states they operate in when implementing the use of automated content removal tools, particularly if there are any considerations of sole reliance on an automated process in the future. However, it was highlighted that it would not be possible to address the issue of false positives and false negatives when there is a lack of universal agreement on what even constitutes 'extremist' content. It was also found that the use of AI in content moderation is limited in effectiveness as it cannot accurately flag certain extremist content, leading to the normalisation of extremist content among social media platforms. It was demonstrated that strict regulatory frameworks on content moderation have created an increasing reliance on automated content removal which, arguably, undermines the basic principles of democracy as social media companies are left to make decisions on what is considered extremist content.

Additionally, there are concerns surrounding violations of the freedom of expression as it limits the users right to free speech by encouraging the removal of legal content. This is further heightened by the lack of transparency and accountability for companies during the decision-making process of automated content removal which, in turn, impacts the appeals process. However, it should be acknowledged that some of these criticisms are inevitable, yet many of them can be prevented using additional or alternative CVE online measures. The development of new regulatory structures, such as 'e-courts', have been proposed which may provide a possible solution as the power of social media platforms over their user's expression would reduce and, consequently, support democracy. De-platforming measures were shown to be a more effective tool to counter violent extremism online, particularly far-right extremists, as it has been shown to effectively reduce the overall reach of far-right extremist narratives. Yet, it was emphasised that more empirical research is needed to determine the extent of the impact that de-platforming measures have had on reducing the exposure of mainstream social media users to other violent extremist groups. Due to the effectiveness of de-platforming measures in CVE, it is important to consider the use of AI in the de-platforming measures as it could arguably increase the capacity and ability to remove extremist groups and limit the reach of their ideologies.

References

Allen, C. (2021, March 29). Extremism in the UK: New definitions threaten human and civil rights. *The Conversation*. https://theconversation.com/extremism-in-the-uk-new-definitions-threaten-human-and-civil-rights-157086

Centre for Data Ethics and Innovation. (2020). *Online targeting: Final report and recommendations*. https://www.gov.uk/government/publications/cdei-review-of-online-targeting/online-targeting-final-report-and-recommendations#fn:189

Centre for Data Ethics and Innovation. (2021). *The role of AI in addressing misinformation on social media platforms*. https://assets.publishing.service.gov.uk/government/uploads/system/uploads/attachment_data/file/1008700/Misinformation_forum_write_up__August_2021__-_web_accessible.pdf

Chaabene, N. E. H. B., Bouzeghoub, A., Guetari, R., & Ghezala, H. H. B. (2021, October). Applying machine learning models for detecting and predicting militant terrorists behaviour in Twitter. In *2021 IEEE international conference on systems, man, and cybernetics (SMC)* (pp. 309–314). IEEE. https://ieeexplore.ieee.org/stamp/stamp.jsp?tp=&arnumber=9659253

Coeckelbergh, M. (2019). Artificial intelligence: Some ethical issues and regulatory challenges. *Technology and Regulation, 2019*, 31–34. https://techreg.org/article/view/10999/11973

Conway, M. (2020). Routing the extreme right: Challenges for social media platforms. *The RUSI Journal, 165*(1), 108–113. https://www.tandfonline.com/doi/pdf/10.1080/03071847.2020.1727157?needAccess=true

Díaz, Á., & Hecht-Felella, L. (2021). *Double standards in social media content moderation*. https://www.brennancenter.org/sites/default/files/2021-08/Double_Standards_Content_Moderation.pdf

Donovan, J., Lewis, B., & Friedberg, B. (2018). Parallel ports. Sociotechnical change from the Alt-Right to Alt-Tech. In M. Fielitz & N. Thurston (Eds.), *Post-digital cultures of the far right* (pp. 49–66). Transcript Verlag. https://www.degruyter.com/document/doi/10.14361/9783839446706-004/html

Duarte, N., Llanso, E., & Loup, A. C. (2018). *Mixed messages? The limits of automated social media content analysis*. https://cdt.org/wp-content/uploads/2017/12/FAT-conference-draft-2018.pdf

Engstrom, E., & Feamster, N. (2017). *The limits of filtering: A look at the functionality and shortcomings of content detection tools*. https://static1.squarespace.com/static/571681753c44d835a440c8b5/t/58d058712994ca536bbfa47a/1490049138881/FilteringPaperWebsite.pdf

Federal Bureau of Investigation. (n.d.). *What is violent extremism?*https://www.fbi.gov/cve508/teen-website/what-is-violent-extremism

Federal Cabinet. (2017). *Declaration on freedom of expression*. https://deklaration-fuer-meinungsfreiheit.de/en/.

Ganesh, B., & Bright, J. (2019). *Extreme digital speech: Contexts, responses, and solutions*. https://www.voxpol.eu/download/vox-pol_publication/DCUJ770-VOX-Extreme-Digital-Speech.pdf

Gillespie, T. (2020). Content moderation, AI, and the question of scale. *Big Data & Society, 7*(2), 1–5. https://journals.sagepub.com/doi/pdf/10.1177/2053951720943234

Global Network Initiative. (2017). *Proposed German legislation threatens free expression around the world*. https://globalnetworkinitiative.org/proposed-german-legislation-threatens-free-expression-around-the-world/

Gorwa, R., Binns, R., & Katzenbach, C. (2020). Algorithmic content moderation: Technical and political challenges in the automation of platform governance. *Big Data & Society, 7*(1), 1–15. https://journals.sagepub.com/doi/full/10.1177/2053951719897945

Guhl, J., Ebner, J., & Rau, J. (2020). *The online ecosystem of the German far-right*. https://www.isdglobal.org/wp-content/uploads/2020/02/ISD-The-Online-Ecosystem-of-the-German-Far-Right-English-Draft-11.pdf

Henschke, A., & Reed, A. (2021). Toward an ethical framework for countering extremist propaganda online. *Studies in Conflict & Terrorism*, 1–18. https://www.tandfonline.com/doi/pdf/10.1080/1057610X.2020.1866744?needAccess=true

HM Government. (2019). *Online harms white paper*. https://dera.ioe.ac.uk/33220/1/Online_Harms_White_Paper.pdf

Home Office. (2015). *Revised prevent duty guidance: For England and Wales*. https://www.gov.uk/government/publications/prevent-duty-guidance/revised-prevent-duty-guidance-for-england-and-wales

Home Office. (2018). *New technology revealed to help fight terrorist content online*. https://www.gov.uk/government/news/new-technology-revealed-to-help-fight-terrorist-content-online

Kundnani, A., & Hayes, B. (2018). *The globalisation of countering violent extremism policies: Undermining human rights, instrumentalising civil society*. https://www.tni.org/files/publication-downloads/the_globalisation_of_countering_violent_extremism_policies.pdf

LaFree, G., & Freilich, J. D. (2019). Government policies for counteracting violent extremism. *Annual Review of Criminology, 2*, 13.1–13.22. https://intranet.swan.ac.uk/webtemp/automailer/e.g.pearson_04112020_e.g.pearson_04112019_Freilich_&_LaFree.pdf

Lee, B. (2020). Neo-Nazis have stolen our memes: Making sense of extreme memes. In M. Littler & B. Lee (Eds.), *Digital extremisms: Readings in violence, radicalisation and extremism in the online space* (pp. 91–108). Palgrave Studies in Cybercrime and Cybersecurity.

Liang, C. S., & Cross, M. J. (2020). White crusade: How to prevent right-wing extremists from exploiting the internet. *Geneva Centre for Security Policy*, (11), 1–27. https://dam.gcsp.ch/files/doc/white-crusade-how-to-prevent-right-wing-extremists-from-exploiting-the-internet

Llansó, E. J. (2020). No amount of "AI" in content moderation will solve filtering's prior-restraint problem. *Big Data & Society, 7*(1), 1–6. https://journals.sagepub.com/doi/pdf/10.1177/2053951720920686

Llansó, E., Van Hoboken, J., Leerssen, P., & Harambam, J. (2020). *Artificial intelligence, content moderation, and freedom of expression*. https://www.ivir.nl/publicaties/download/AI-Llanso-Van-Hoboken-Feb-2020.pdf

Lowe, D. (2017). Prevent strategies: The problems associated in defining extremism: The case of the United Kingdom. *Studies in Conflict & Terrorism, 40*(11), 917–933. https://www.tandfonline.com/doi/pdf/10.1080/1057610X.2016.1253941?needAccess=true

Miller, J. (2017, June 30). Germany votes for 50m euro social media fines. *British Broadcasting Corporation News*. https://www.bbc.co.uk/news/technology-40444354

Neumann, P. R. (2013). The trouble with radicalization. *International Affairs, 89*(4), 873–893. https://www.jstor.org/stable/pdf/23479398.pdf?ab_segments=0%2Fbasic_search_solr_cloud%2Fcontrol&refreqid=fastly-default%3Aa01b19ab612357c46a149138dd388925

Nouri, L., Lorenzo-Dus, N., & Watkin, A. L. (2019). *Following the whack-a-mole: Britain First's visual strategy from Facebook to Gab*. https://static.rusi.org/20190704_grntt_paper_4.pdf

Nouri, L., Lorenzo-Dus, N., & Watkin, A. L. (2021). Impacts of radical right groups' movements across social media platforms–a case study of changes to Britain First's visual strategy in its removal from Facebook to Gab. *Studies in Conflict & Terrorism*, 1–27. https://doi.org/10.1080/1057610X.2020.1866737

Ofcom. (2019). *Use of AI in online content moderation*.https://www.ofcom.org.uk/__data/assets/pdf_file/0028/157249/cambridge-consultants-ai-content-moderation.pdf

Parliamentary Office of Science and Technology. (2020). *Online extremism*. https://researchbriefings.files.parliament.uk/documents/POST-PN-0622/POST-PN-0622.pdf

Piazza, J. A., & Guler, A. (2019). *The online caliphate: Internet usage and ISIS support in the Arab world*. https://www.tandfonline.com/doi/pdf/10.1080/09546553.2019.1606801?needAccess=true

Rogers, R. (2020). Deplatforming: Following extreme Internet celebrities to Telegram and alternative social media. *European Journal of Communication, 35*(3), 213–229. https://journals.sagepub.com/doi/full/10.1177/0267323120922066

Schmid, A. P. (2014). *Violent and non-violent extremism: Two sides of the same coin*. https://opev.org/wp-content/uploads/2019/10/Violent-and-Non-Violent-Extremism-Alex-P.-Schmid.pdf

Selim, G. (2016). Approaches for countering violent extremism at home and abroad. *The Annals of the American Academy of Political and Social Science, 668*(1), 94–101. https://journals.sagepub.com/doi/pdf/10.1177/0002716216672866

Stephens, W., Sieckelinck, S., & Boutellier, H. (2018). Preventing violent extremism: A review of the literature. *Studies in Conflict & Terrorism*, 1–16. https://www.tandfonline.com/doi/full/10.1080/1057610X.2018.1543144

Tworek, H., & Leerssen, P. (2019). *An analysis of Germany's NetzDG law*. https://cdn.annen berg-publicpolicycenter.org/wp-content/uploads/2020/06/NetzDG_TWG_Tworek_April_2019.pdf

United Nations Office of Counter-Terrorism, & United Nations Interregional Crime and Justice. (2021). *Countering terrorism online with artificial intelligence*. https://www.un.org/counterterrorism/sites/www.un.org.counterterrorism/files/countering-terrorism-online-with-ai-uncct-unicri-report-web.pdf

United Nations Office on Drugs and Crime. (2012). *The use of the Internet for terrorist purposes*. https://www.unodc.org/documents/frontpage/Use_of_Internet_for_Terrorist_Purposes.pdf

van der Vegt, I., Gill, P., Macdonald, S., & Kleinberg, B. (2019). *Shedding light on terrorist and extremist content removal*. https://rusi.org/explore-our-research/publications/special-resources/shedding-light-on-terrorist-and-extremist-content-removal

Vincent, C., & Hunter-Henin, M. (2018, February 10). The trouble with teaching 'British values' in school. *Independent*. https://www.independent.co.uk/news/education/british-values-education-what-schools-teach-extremism-culture-how-to-teachers-lessons-a8200351.html

Wakefield, J. (2021, May 12). Government lays out plans to protect users online. *British Broadcasting Corporation News*. https://www.bbc.co.uk/news/technology-57071977

Weimann, G., & Am, A. B. (2020). Digital dog whistles: The new online language of extremism. *International Journal of Security Studies, 2*(1), 4. https://digitalcommons.northgeorgia.edu/cgi/viewcontent.cgi?article=1030&context=ijoss

West, L. J. (2021). Counter-terrorism, social media and the regulation of extremist content. In S. Miller, A. Henschke, & J. Feltes (Eds.), *Counter-terrorism: The ethical issues* (pp. 116–128). Edward Elgar Publishing. https://www.elgaronline.com/view/edcoll/9781800373068/9781800373068.00016.xml

Wojcicki, S. (2017, December 5). Expanding our work against abuse of our platform. *Youtube Official Blog*. https://blog.youtube/news-and-events/expanding-our-work-against-abuse-of-our/

A Critical Analysis into the Beneficial and Malicious Utilisations of Artificial Intelligence

Megan Thomas-Evans

Abstract Artificial intelligence (AI) has grown exponentially over the past several years, due to technological advancements evolving at an ever-growing rate. Despite the many benefits societies see in AI through the Internet of Things (IoT), there is a malicious layer to the capabilities of AI technologies. By conducting an analysis of the existing literature, this chapter aims to investigate the use and malicious use of AI within businesses and security contexts. Recognising that malicious uses of AI will not be eradicated entirely, developing AI technologies could outweigh the malicious uses in the future.

Keywords Artificial intelligence · Machine learning · Natural language processing · IoT · Counterterrorism · Cybercrime

1 Introduction

1.1 Background and Definitions

Within the field of AI, many definitions can be considered. One early sourced definition states AI is 'the part of computer science concerned with designing intelligent computer systems, that is, systems that exhibit characteristics we associate with intelligence in human behaviour – understanding language, learning, reasoning, solving problems, and so on' (Feigenbaum et al., 1981). This dated definition can still be acknowledged in the current day as more recent definitions also highlight the term 'imitation of human behaviour' (Kok et al., 2009). In addition to previous definitions, AI can be known as the capability of a machine or device to acquire human experiences via a cognitive process (Tyagi, 2020a). The first acknowledgement of AI occurred in the late 1930s by Turing, as he described computer intelligence as a machine that can learn from experience. Turing (1948)

M. Thomas-Evans (✉)
Independent Researcher, Wales, UK

R. Montasari (ed.), *Artificial Intelligence and National Security*,
https://doi.org/10.1007/978-3-031-06709-9_5

labelled this as 'intelligent machinery' which first trialled using games such as chess and checkers by IBM. The AI ability then expanded to reasoning logically (Newell & Shaw, 1957), programming languages (Newell et al., 1958; McCarthy, 1960) and expert systems. AI also evolved in the health industry, originally investigating how the human brain works at the neural level such as retaining knowledge and memory (McCulloch & Pitts, 1943; Farley & Clark, 1954). There is a broad range of techniques that fall into the territory of AI. These include ML, NLP, robotic process automation (RPA), bias, and information processing language (IPL), which will be discussed throughout this chapter. Feigenbaum et al. (1981) also mention that the most important tools within AI are the programming languages that are used when conceiving and implementing the programs. List-processing language, otherwise known as information processing language (IPL), was one of the earliest language programming developed by Newell and Shaw (1957). IPL was developed consisting of symbols rather than numbers, which led to list processing to be created to form associations.

1.1.1 Machine Learning

ML technology is the technique that enables computers to learn without being programmed (Expert.ai, 2017). This is completed via ML algorithms continuously being learned on technology systems. Applications of ML are being utilised in everyday life without recognising it is present (Dar, 2019), in applications such as computer vision, speech recognition, NLP, and other applications (Jordan & Mitchell, 2015). One example of a daily life use of ML is commute estimation (Tyagi, 2020b). Google Maps uses location data from smartphones to identify traffic at any time, along with users reporting road-traffic accidents and road works. The cumulative information given to the ML technology allows increased efficiency for users, such as providing the quickest commute route. Regardless of ML managing everyday tasks for users, several security risks have risen within society as it remains a young field of research. Jordan and Mitchell (2015) suggest how ML raises questions regarding which of its potential uses society should support and oppose. The substantial rise in ML technologies over the past few years has resulted in record numbers of personal data collection, which leads to privacy issues. The relationship between online sources, online medical data, and location data can pose a threat to personal privacy if society intervenes. Jordan and Mitchell (2015) state that although the data is readily available online within company databases, laws are not currently in place for society to benefit from them. This displays potential misuse of AI technologies, despite the many beneficial aspects it holds at face value.

1.1.2 Natural Language Processing

NLP is another branch of AI which plays an important role in everyday life within technology. NLP is a domain of AI that aids communication between a computer system and a human by natural language (SAS, n.d.). This technique enables a computer to search, read, analyse, and understand data by imitating human language. For meaningful data to be extracted, computers are taught to do this via using NLP libraries (Chowdhury, 2003). NLP is beneficial to users as accuracy and efficiency of files are increased due to the automatic detection of words, making it an easier task for users to look at important emails (Belsare & Bhate, 2020). In customer service, NLP is used to operate live chat conversations to filter customers to correct departments which will find quicker solutions (Lester et al., 2004). However, this can be a challenge in some customer queries as the customer may type text that is not detected by the pre-programmed word set. Chatbots can communicate and ask questions to numerous customers simultaneously, whereas a human operator would not be able to; 86% of customers prefer to communicate with a human to receive an accurate response (Press, 2019). Another example of NLP is text translation, recently implemented by Google and Samsung phones. This operates by hovering the device's camera over written text and will automatically translate to the desired language. In a terrorism context, Twitter use the NLP technique to filter terrorist language from various tweets. This poses an advantage towards content moderation as a human moderator would be unable to remove terrorist material if there were hundreds of Tweets to filter through. The automaticity of NLP provides social media platforms with the ability to remove terrorist content and hate speech instantly, decreasing the number of users who could become radicalised from this content being present (Tyagi, 2020a).

1.2 Research Questions

The following research questions have been developed based on the background research to address the overarching statement of this chapter: 'A Critical Analysis into the Benefits and Malicious Uses of Artificial Intelligence".

RQ1. How is AI used maliciously?
RQ2. What are the beneficial uses of AI?
RQ3. What are the challenges that AI is confronted with?

1.3 The Structure of This Chapter

The remainder of the chapter is structured as follows. Section 2 will cover RQ1. In this section, the ways in which AI is used maliciously in various contexts will be discussed along with the disadvantages of AI specifically within law enforcement. Several aspects of AI will be evaluated such as the malicious uses in physical devices and within cybercrimes. Section 3 addresses RQ2 focusing on the advantages of AI. This section will discuss the beneficial aspects of the emergence of AI, focusing on how it can be advantageous to law enforcement and businesses. Section 4 answers RQ3 providing an evaluation of the challenges that AI faces such as content moderation, ML, and ethical and legal challenges. Finally, this chapter is concluded in Sect. 5.

2 The Malicious Use of Artificial Intelligence

2.1 Unmanned Aerial Vehicle

One malicious way an individual or group could use AI maliciously is through the physical ability of using unmanned aerial vehicles (UAVs) commonly known as drones. UAVs can be remotely controlled or can fly autonomously controlled by a software flight plan (Tlight, 2020). More commonly, drones are piloted by an individual from the ground using a navigational device and GPS. Traditionally, drones were utilised for military operations, but, more recently, have become popular for personal use (Tlight, 2020). Drones have increasingly become an issue in UK prisons to deliver drugs, weapons, mobile phones, and other contraband (Webster, 2017). In 2018, seven individuals were jailed for flying more than half a million pounds worth of drugs into prison via drones (BBC, 2018a). Prior to this, a British prison became the world's first to design a system which prevents drones from flying over perimeter walls to eliminate contraband coming into jails. An electronic disruptor shield is used to block drone's frequencies and control protocols. The drone defence system, 'Sky Fence' does not damage the drone or hack any intelligence network. However, it forces the operator's device screen to go black and be returned to its original location when the drone reaches the perimeter (Webster, 2017).

The future of AI holds the potential for future terrorist attacks using drones. Tactics and techniques of terrorism are evolving with current and future technology posing a threat to organisations, infrastructures, citizens, and cities across the world (Pledger, 2021). Drones provide the ability of not requiring human interaction in a proximity, which can enable terrorists to conduct attacks without being present and without suicide. They also provide solo or lone terrorists with the ability of conducting multiple terrorist attacks simultaneously via AI technologies such as automation. Past terrorist attacks have included legally purchased materials and weapons, along with individuals that would sacrifice their own life during the attack. Pledger (2021) states that, between 1994 and 2018, over 14 organised terrorist attacks occurred with the use of drones. One example of this occurred in 2013 by Al-Qaeda in Pakistan, even though it was stopped by law enforcement. An example of how AI has been used maliciously in a real-life scenario involves the Venezuelan president. In 2018, two drones detonated explosives in Caracas, Venezuela, where President Maduro was at a military event addressing the 81st anniversary of the Bolivarian National Guard. Seven soldiers were injured as a result; however, the outcome could have been detrimental to many lives. The BBC (2018b) reported 'Soldiers in T-Shirts' had admitted that they had been behind the attack but had failed due to their drones being shot down by the military. However, there was no evidence supporting the 'Soldiers in T-Shirts' claim. As this follows the previous helicopter attack on Venezuela's Supreme Court when they were victims to grenades in June 2017. The more recent use of AI shows how individuals are adopting this style of violence in the political world. Prior to these violent attempts, in 2013 a device was captured hovering above a crowd when the German Chancellor, Angela Merkel, was giving a speech. The operator was ordered to lower the drone, which concluded the incident. Furthermore, drones have also been used in the UK to disrupt airport operations. At a busy time of year for the aviation industry, in December 2018, Gatwick airport was subject to drone identification. The threat that this caused prompted terror within the vicinity due to the interference it could have had upon aircrafts or possible risk of explosions. The three-day disruption cost the airport £1.4 m (Topham, 2019) along with costing airlines an estimated £50 m (Calder, 2019). The financial implications this event presented to Gatwick airport and airlines reveal the extent of detrimental damage drones can cause.

2.2 3D Printing

Another malicious use of AI involves 3D printing, the exploitation of which provides terrorists with the ability to manufacture and produce dangerous weapons such as homemade guns. The Halle-attack in Germany in 2019 composed of guns created by 3D printers. It is stated how accessible and easy it was to develop these weapons: 'you need a weekend worth of time and $50 for the materials' (van der Veer, 2019). The gun became jammed in the attack, proving to be unsuccessful. As technology is improving rapidly, researchers warn this could lead to a large risk posed by 3D printing techniques as the quality of 3D printing increases (van der Veer, 2019), especially in countries with high gun control (Mapua, 2019). Improvised firearms by violent non-state actors introduces the emergence of improving manufacturing technologies (Veilleux-Lepage, 2021). 3D printing enthusiasts have taken not only to producing weapons but also to creating the traditional 'how to' guides for like-minded individuals. Within terrorist social media platforms, originally 'how to' guides were published and distributed to create homemade bombs and explosives. As technological capabilities improve, the 'how to' guides are also advancing by including complex instructions such as how to operate 3D printing machinery and create more powerful weapons. Police recently identified terrorism manuals in a Spanish shop after dismantling an illegal workshop that was producing 3D-printed weapons (BBC, 2021). Law enforcements should prioritise destroying 3D printing workshops with the aim to eradicate the manufacturing process of weapons. Although it will not stop the use of printing 3D weapons entirely, it will decrease the amount vastly.

2.3 Distributed Denial of Service (DDOS) Attacks

There are alternative ways in which AI can be used maliciously in the digital world compared to physical attributes. AI can also be used maliciously in the context of cybercrime. This includes its malicious use in cybercrimes such as hacking, phishing, and Distributed Denial of Service (DDoS) attacks. DDoS attacks are malicious attempts to cause a disturbance to the normal traffic of a network, service, or server by overwhelming its infrastructure with a surge of Internet traffic (Cloudflare, 2021). Cybercriminals commit DDoS attacks to remotely control a device to server and cause harm, and are often used as a tool of extortion, blackmail, politics, and business conflict (GenieNetworks, 2019). Sophisticated technologies such as AI and ML have become integrated in DDoS attacks and other cybercrimes based on algorithms and techniques to target specific servers. AI facilitates DDoS attacks through operations such as automation. The automatic spreading of malware (malicious software) without human intervention increasingly became present during the evolution of DDoS attacks (Conran, 2018). Whereas cybercriminals can use AI maliciously to conduct DDoS attacks, AI, itself, can be utilised to detect and prevent such attacks (Glăvan et al., 2019).

2.4 Phishing

Phishing is another cybercrime that regularly affects citizens' personal data. Phishing can be defined as the fraudulent practice of sending emails imitating businesses to deceive individuals to reveal personal information (Merriam-Webster, n.d.). This personal data can include passwords, bank details, and email addresses. Phishing attacks can also be carried out through email, text messages, telephone calls as well as social media. Phishing emails will include a link, encouraging the receiver to click on it with a monetary reward, or to falsely update personal details. AI facilitates phishing by automatically downloading malware onto the user's device or direct them to a malicious website causing a threat to IT systems. Due to the automation facility, phishing attempts can reach thousands of users simultaneously, including organisations of any size. Malware detection tools must stay up to date with technology to remain effective and secure as possible (Vigna, n.d.). Hackers are developing more innovative ways to deliver phishing messages as more people are recognising fraudulent emails. A benefit of AI within phishing scams features NLP. Spam emails are recognised by pre-programmed algorithms through NLP and are directed to junk boxes. However, phishing emails can seem realistic when they bypass the junk box by the sender avoiding key terms, and directly enter the inbox. Implications of this could result in individuals being scammed of thousands of pounds or damage to infrastructures. As around half of cyberattacks in the UK involve phishing (Lavion et al., 2018), this encourages NLP to stay up to date with recent terminology, to divert scams to junk boxes to alert caution.

3 The Beneficial Uses of Artificial Intelligence

3.1 Efficiency and Automation

One major beneficial aspect of AI is the efficiency it holds. The use of AI enables increased efficiency in many areas. AI is valuable in improving data analysis speed and increases reporting time (Gardner, 2019). This remains beneficial within data-analysis roles and businesses as the efficiency of the technology can prevent human error from occurring in the workplace. Analysing via AI enables data to be processed and analysed more accurately (Longoni et al., 2019), with larger data sets (Noorbakhsh-Sabet et al., 2019). This is particularly highlighted in healthcare as AI can provide decision support systems at larger scales than what a human nurse or doctor could deal with independently (Noorbakhsh-Sabet et al., 2019). AI systems are also able to work 24/7, which would be costly for a business to pay employees to work throughout the day and night conducting the same role when less work would be completed. Furthermore, humans can fatigue during a work shift, leaving their performance to decline (Åkerstedt & Wright, 2009). Another benefit of AI is automation. Platforms such as Gmail and Outlook use AI to give automated reminders to users about emails that may need action. Due to the COVID-19 pandemic, many organisations were forced to work from home during the pandemic. Workforces have implemented applications such as Microsoft Teams and Outlook to their everyday communication. Within Microsoft, AI has aided efficiency in the working day of employees via Cortana, which is an AI-enabled virtual assistant. Cortana sends personalised briefing emails to users including tasks and commitments recognised by previous emails (Microsoft, 2021). This is very beneficial to individuals who deal with multiple emails every day and those with a high workload. It provides them with increased organisation, which, therefore, increases efficiency. Automation tools are also beneficial within the context of security.

Autonomous and intelligent systems have previously been utilised in military technology, and evolvement in ML and AI have provided vast advancements in automation within warfare. Allen and Chan (2017) suggest ML technology can enable a high level of automation in heavy-duty tasks such as satellite imagery analysis and cyber defence. The future for military security within AI suggests that tasks previously incapable of being automated will now become automated (Felten, 2016). This will decrease the need for human interaction in the cyber domain. As automated processes can be conducted more efficiently and more accurately than humans, this allows staff to remain safe and not at potential harm. Another tool of AI that is adopted to protect national security is ML in content moderation. This takes an approach from the social media perspective, to protect users online from online radicalisation and viewing malicious content. One of the major issues social media moderators are trying to overcome is the spread of terrorist propaganda and hate speech online. Many companies are turning to AI to tackle this issue, by using automated hash-matching and predictive ML techniques (Gorwa et al., 2020).

The use of algorithmic programming to automate the removal of hate speech and malicious content can help to reduce the number of users subject to viewing violent content online. In turn, this will ensure that these individuals will be less likely to be radicalised or harmed online.

3.2 Business

AI also has many benefits within business to reduce operational costs, increase efficiency, and improve customer service (NI Business Info, n.d.). Companies rely on quantitative data to gain valuable insights on strategies that can increase businesses' growth and profit (Business World, 2020). Using AI, companies can engage customers and automate business processes, while reducing operational expenses. As discussed previously, the use of chatbots have significantly increased in recent years within customer service (Hill et al., 2015). Chatbot is an AI software that manages conversations with customers in natural language (Ho, 2021). A chatbot is operated by NLP using the responses from human language (Thomas, 2016). Chatbots are considered more convenient for customers shopping online who seek customer support, payment queries, and alternative inquiries without holding for long periods of time over the phone (Ho, 2021). As many internet activities are becoming an instant feature, such as same-day deliveries and instant bank transfers via online banking, it could be deemed imperative to provide instant customer service to maintain customer satisfaction. NLP is used to operate live chat conversations to filter customers to correct departments which will find quicker solutions (Lester et al., 2004). However, due to the chatbot being programmed with pre-determined responses, the chatbot cannot provide flexible answers when the question is not recognised in the system by algorithms (Jeong & Seo, 2019). Although chatbots can communicate and ask questions to numerous customers simultaneously, 86% of customers prefer to communicate with a human to receive an accurate response (Press, 2019). To overcome this, chatbot 'training' is being implemented to train more algorithms to answer more questions proposed by customers (Kvale et al., 2019). It is believed that in the next few years, chatbots will replace customer service assistants due to the improvements in AI technology, particularly in NPL.

3.3 Prisons

In addition, AI improves many sectors within prisons such as security, maintenance, education, and surveillance. The ways in which AI can provide prisons with improved surveillance is through the monitoring of inmate phone calls, the ability to gain knowledge of criminal activity, learn about potential suicides (Weiss, 2019), and recognise violence occurring (Kanowitz, 2021). Within surveillance in prisons, several branches of AI are utilised. These include: speech recognition, semantic analytics, and ML. Inmate phone calls are monitored by AI companies, who notify law enforcement when suspicious language is recognised. This language is recognised via cloud-based NLP which is used to create a customised lexicon determined by inmates' language based on key terms, code words, and slang (Kanowitz, 2021). This is advantageous to law enforcement as human moderation is not required, which would be seen as an invasion of privacy by prison inmates if this was needed. However, as words are pre-selected to be detected by the language recognition system, new terminology and slang could be created by inmates communicating over the phone, which would bypass the speech recognition technology. This requires the customised lexicon to be updated on a regular basis. This could cause a hindrance to AI technology companies providing the service to the prison. Another drawback for this technology is regarding the inmates that have not yet been convicted of their crime. They remain under surveillance which could be considered an invasion of privacy if they are proven not guilty. However, it is imperative that the prison remains secure, as unsentenced inmates may still cause a threat to security. Prisons are also using AI to decrease the amount of crime occurring in prisons. The less crime taking place inside prisons results in fewer crimes being committed outside when prisoners are released (Gov.uk, 2018).

New implementations are aiming to achieve this using new technological solutions. Outside of the UK, specifically Finland, sees a rapid advance in technological designs such as AI. In October 2018, the Smart Prison Project was developed aiming for a new prison concept that utilises digital services for rehabilitation, education, and reintegration into society by learning about AI on the smart systems (Järveläinen & Rantanen, 2020). The Hämeenlinna women's prison had completed its integration of digital services in November 2020, which included laptops with smart systems being installed into each cell (Rantanen et al., 2021). Each laptop consisted of the ability to send messages, requests, and make video calls to members of staff at the prison and access health care (Puolakka, 2020). This limits the amount of time inmates require assistance from prison guards, increasing independence. The laptop also included information on AI for the prisoners to learn about the subject, increasing their knowledge and employability for when they leave prison.

3.4 Counterterrorism

AI can also help to contribute to counterterrorism operations. Intelligence has employed many AI technologies such as ML and NLP to tackle terrorism. AI is advantageous to intelligence due to the large data sets available for analysis which would be difficult for human evaluation. Face recognition has also been adopted in security measures to identify individuals who are deemed a threat to the public via CCTV (McKendrick, 2019). Face recognition uses deep learning algorithms via ML to compare a live captured image to the stored face image to verify an individual's identity (Gillis, 2019). It has been proven to be the least intrusive but fastest form of biometric verification (Garg, 2020). Deep learning algorithms provide CCTV with the ability to identify individuals known as missing persons, individuals in public media images, and verify mugshots compared against a database (Swindler, 2018). Shopping centres, airports, and sports stadiums are a few examples of where face recognition is used to prevent shoplifters and terrorists. Swindler (2018) suggests analytics can be produced from facial recognition software via CCTV to prepare security of the most popular times and locations these offences occur. Employers can then distribute more security around this time to deter criminals. If crime occurs in a particular store within a shopping centre, the retailers could distribute loss prevention staff to decrease the likelihood of crime. On the other hand, there are some ethical implications associated with facial recognition such as the issues of bias. Facial recognition algorithms retain high classification accuracy at over 90%. However, a study (Klare et al., 2012) has found consistent errors in algorithms related to the facial recognition of those who are female and Black, and are aged 18–30 years old. Klare et al. (2012) suggest that by focusing on improving recognition accuracy within AI technologies in neglected demographics, researchers will be able to reduce error rates, and law enforcements would be more successful in identifying correct individuals on facial recognition systems.

Law enforcement and counterterrorism departments are constantly seeking for advancing technologies to help them be one step ahead of criminals and cyber criminals (Devi & Sairam, 2018). Counterterrorism approaches involve biometric modalities such as face recognition due to its low cost, contactless ability, and without the consent of individuals (Ghalleb & Amara, 2020). Additionally, face recognition has many government documents to compare to such as passports, driving licenses, and ID cards. However, there are many factors which hinder face recognition in real life scenarios (Gonzalez, 2013). Face recognition remains an active area of research due to its inability to recognise individuals in different lighting, distorted background, distance from sensor, and emotion along with other factors mentioned by Ghalleb and Amara (2020).

4 The Challenges Faced by Artificial Intelligence

4.1 Technical Challenges

4.1.1 Content Moderation

Despite the many beneficial aspects ML holds, there are several challenges it faces within content moderation. As previously discussed, social media platforms are challenged with moderating hate speech and terrorist propaganda being posted online. Although the basic capabilities of AI and specifically ML can remove and block posts to decrease the number of users seeing terrorist propaganda or offensive material, AI technologies require more tailored algorithms to become more accurate in removing the harmful posts. Content can appear in many different formats such as text, image, video, or audio, making it difficult for human moderators to detect (Llansó et al., 2020). There exists some difficulties posed by AI in moderating content as well as humans. One example of this relates to multilingual text. Multilingual text cannot be automatically translated to another language as tools are designed to detect 'toxic' comments in one language (Tellez et al., 2017). In relation to image analysis, it is conducted by automated image detection and identification tools to recognise previously identified content along with more complex tools such as discovering close features (Llansó et al., 2020). Llansó et al. (2020) describe hash values as unique numerical values generated via specific algorithms on a file. Factors that the function evaluates include image dimensions and colours of pixels. However, individuals who attempt to avoid detection by moderators, change these factors such as altering a colour within a pixel, or cropping the image; in turn, this alters the hash value of the file, allowing it to be undetected and therefore remains online (Locascio, 2018). Perceptual hashing is more resilient in detecting images that are required to be removed from social media sites. This is due to the fact that the technique calculates hashes based on relationships within pixels and accounts for minor discrepancies in the final hash (Segers, 2014).

Another image analysis technique involves the method of ML from the discipline of computer vision. Computer vision enables mobile devices to extract the meaning of the image displayed through onboard sensor cameras (Battiato et al., 2012). Specific features that can be recognised by algorithms include searching for weapons, symbols, face detection, or logos. This is used in attempting to remove terrorist content as terrorist group logos can be automatically detected by this technique along with images of weapons. However, ML tools that hold predetermined images such as terrorist group logos must be able to identify several variations with different orientation, lighting, and resolutions (Llansó et al., 2020). For example, if the logo is rotated 90°, blurred, and darker than the predetermined image, it may not be recognised by the image analysis. Optical character recognition (OCR) is used to detect text within images; however, a language-correction phase based on NLP is used to minimise the potential for incorrect readings to evaluate the meaning of text accurately (Apple, n.d.).

With regards to automation, one way in which ML techniques are used in automation is removal of online text. This is conducted through keyword filtering which automatically blocks posts or access to websites when predetermined key words are detected (Llansó et al., 2020). However, this can be deemed underinclusive and narrow since posts with a positive intention could be blocked as the system has detected it in a malicious way. Therefore, NLP is utilised to provide a more comprehensive analysis of the text (CDT, 2014). NLP tools are trained to predict whether the text is communicating a positive or negative emotion through sentiment analysis which will classify it in either 'hate speech' or 'not hate speech'. Recent research experiments emojis in sentiment-analysis since they are seen as being involved within the modern language online (Felbo et al., 2017). However, NLP perform best when used with data it was trained on. This is because it could not be generalised to other settings such as a change in culture, language, or interest groups (Llansó et al., 2020).

In respect of ML methods, ML is utilised in two ways during content moderation: supervised and unsupervised learning. Supervised learning can be described by its use of labelled datasets to train algorithms that predict outcomes accurately (IBM Cloud Education, 2020). Human moderators label a group of posts as 'hate speech' or 'not hate speech' which enable the ML model to identify features of the dataset and categorise future posts. However, due to the posts being manually inputted by human moderators, this can cause bias and errors entering the model, causing low intercoder reliability (Amini et al., 2019). For example, one human moderator could label a term 'racist', whereas another moderator does not agree; this has future implications for the model to become inconsistent and, therefore, not to work towards its purpose. On the other hand, unsupervised learning involves training a model based on an unlabelled dataset (Llansó et al., 2020). This model learns to recognise underlying patterns within the data, including detecting pairs of words that are often present together. These pairs will then become the 'labelled' training data that can be assessed whether to remove or keep on the platform. A challenge that unsupervised ML faces is overfitting. Domingos (2012) recommends cross-validation, regularisation, and Chi-squared testing for adjusting an unsupervised learning algorithm to avoid overfitting. Overfitting could occur when paired words are repeatedly recognised but are unrelated to malicious posts. The paired words could be deemed offensive when placed together, but the algorithms ignore contextual value which is unharmful when placed in a sentence.

4.1.2 Facial Recognition Bias

Another technical AI challenge within includes the issue of potential discrimination presented by facial recognition software. This is derived from emotional analysis from two different facial recognition services, 'Face', and 'Microsoft's Face API'. Rhue (2018) found both services interpret Black basketball players producing more negative emotions than white players along with registering angrier faces in Black players compared to white players. Microsoft registered a 'contempt' emotion rather than 'anger' and interprets Black players as more contemptuous when their facial expressions are ambiguous (Rhue, 2018). As stated previously, discrepancies have also been found (Klare et al., 2012) in facial recognition when detecting Black faces. As these research papers are six years apart, it suggests not much improvement has been made in racial diversity within AI facial recognition technologies. For AI facial recognition software to be considered accurate, algorithms for all races, ethnicities, genders, and ages must be researched and tested more. This will ensure to deliver a higher level of security to the public and eliminate racial bias in AI.

4.2 Legal Challenges

4.2.1 Freedom of Expression

Content moderation also poses a threat to freedom of speech on social media platforms. False negatives and false positives can have implications when removed on online social media platforms. False positives being removed from social media sites can be deemed a burden on individuals' rights to freedom of expression. In contrast, false negatives allow the hate speech, offensive material, or terrorist propaganda to remain on sites, causing a disturbance online to many groups. One issue that social media platforms face when dealing with content moderation regarding freedom of speech is a difference in regulations between the platforms. As the decisions are delegated to private sector entities, policies differ, rendering it difficult to address to users (Karanicolas, 2021). More sophisticated AI systems within online social media platforms may be the result of pressure from governments to remove harmful or illegal content. This pressure may cause vast amounts of content to be removed with no human intervention, resulting in going beyond the limitations on freedom of expression (Hovland & Seetharaman, 2016). On the subject of privacy, AI and the IoT are known to store vast amounts of data without explicit consent from users. As the IoT are ubiquitous in everyday lives, it is important to educate users about the type of information they share about themselves and how much is necessary. Facial recognition systems also present legal challenges. It has been stated that ML facial recognition systems have been able to identify approximately 69% of covered faces (by hats and scarves) during protests (CB Insights, 2017). This can be deemed a breach of privacy as it is assumed that individuals who cover their faces would not like to be identified. However, from a national security and counterterrorism point of view, it is important for public safety to identify potential criminals in such events. This is supported by the context of law enforcement, as facial recognition is allowed for the police to identify someone without reasonable suspicion (Privacy International and ARTICLE 19, 2018).

4.3 Ethical Challenges

Ethics are a broad and integral part of research and must be recognised to support the researcher and participants while protecting personal welfare and personal data. Ethics are involved in AI as a system of moral principles intended to inform the use of AI technology responsibly by using integrated techniques (Lawton & Wigmore, 2021). As AI is designed to replicate humans, the AI systems must replicate the ethics and protection a human researcher would deliver in the real world. Without this, there is a chance of AI systems being poorly designed and having destructive consequences on personal harm, data, and infrastructures. Having ethical AI within businesses reassures customers and enables customer bases to grow within companies. This will uphold customer values allowing customers and the wider audience to perceive the brand in a positive light. Ethical AI also motivates the employees to be proud of the company they work for as they know that neither they nor customers will be harmed.

Lawton and Wigmore (2021) state that, according to Shepherd, there are three key areas which need addressing under ethical AI. These include policy, education, and technology. Ethical AI needs to understand various policies to address legal issues when a mistake is made. Policy protects companies and stakeholders, and they are implemented in the company's own code of conduct. Shepherd states that, along with employees being aware of the policies, consumers need to stay educated with up to date policies to control their data to the best of their ability. As AI technologies are based on automatic processes, when a problem arises, human interaction will be needed. This would require investigation through a complex chain of algorithmic systems to identify the problem. Companies who are involved with AI systems need to have extensive reasoning. The content human moderators are required to analyse can be emotionally and psychologically damaging (Li et al., 2019). As human content moderators see terrorist material, assaults, and other violent images, this can cause a detrimental impact on their mental health. This is one reason for AI algorithms to become more accurate in removing harmful online content to protect the well-being of human moderators. Another ethical challenge of AI is responsibility when issues occur. Destructive consequences can occur through the fault of AI systems, and society still struggle to identify who is responsible when issues arise. Due to the possibility of loss in capital or life, or risk to health, it is important that companies liaise together to identify who would be at fault prior to designing AI systems. When fatalities or injuries occur, there is an argument whether it is the user's fault for not providing human intervention or if it is the fault of the AI technology system.

5 Conclusion

Through a critical analysis, the malicious and beneficial uses of AI along with challenges of AI were evaluated. The literature found several ways in which AI can be used maliciously through cyberattacks such as phishing and DDoS attacks, and alternatively through physical aspects such as drones and 3D printing. While these examples were discussed in detail, it was suggested that malicious uses are not limited to the physical devices and cybercrimes. The key benefits of AI emphasised the ability of automation and efficiency it provided to businesses, counterterrorism, prisons, and law enforcement. This chapter revealed how AI could replace human intervention to prioritise the safety and well-being of employees in various occupations. As well as critically discussing the many benefits of AI, a comprehensive analysis was provided associated with the technical, legal, and ethical challenges presented by AI. The reoccurring theme of content moderation was demonstrated throughout this study. Therefore, improvements to ML technologies are required to address the issues related to content moderation. This is imperative to decrease the psychological and emotional harm that it poses to human content moderators.

In relation to future research, it is suggested that an empirical research study be conducted on how people perceive AI technologies. As these technologies are advancing constantly, it can be deemed difficult for the public to remain up to date with evolving technology. AI technologies are involved in everyday life through smart devices, and it would be interesting to view if people were aware of how AI is used maliciously in the world around us. This will aid individuals in understanding both the benefits and risks that AI presents. Any future empirical research should include a broad participation sample to gain the perspective of all ages to generalise findings to the wider population. Furthermore, improvements must be made to facial recognition technologies and AI algorithms within automated content removal. Accuracies in both are limited even though they can pose a threat to national security. Facial recognition must improve its accuracy in detecting emotions in different ethnic groups before it can be expected to be accurate within the notion of national security. It is also imperative that automated content removal improve its accuracy to protect the well-being of human social media platform content moderators along with social media users.

References

Åkerstedt, T., & Wright, K. P. (2009). Sleep loss and fatigue in shift work and shift work disorder. *Sleep Medicine Clinics, 4*(2), 257–271.

Allen, G., & Chan, T. (2017). *Artificial intelligence and national security*. Belfer Center for Science and International Affairs.

Amini, A., Soleimany, A. P., Schwarting, W., Bhatia, S. N., & Rus, D. (2019). Uncovering and mitigating algorithmic bias through learned latent structure. In *Proceedings of the 2019 AAAI/ACM Conference on AI, Ethics, and Society* (pp. 289–295).

Apple. (n.d.). *Recognizing Text in Images*. Available at: https://developer.apple.com/documentation/vision/recognizing_text_in_images#see-also. Accessed March 08, 2022.

Battiato, S., Farinella, G. M., Messina, E., Puglisi, G., Ravì, D., Capra, A., & Tomaselli, V. (2012). On the performances of computer vision algorithms on mobile platforms. In *Digital Photography VIII* (pp. 8299–82990L). International Society for Optics and Photonics.

BBC. (2021). *Spain dismantles workshop making 3D-printed weapons*. Available at: https://www.bbc.co.uk/news/world-europe-56798743. Accessed March 08, 2022.

BBC. (2018a). Gang who flew drones carrying drugs into prisons jailed. *BBC*. Available at: https://www.bbc.co.uk/news/uk-england-45980560. Accessed March 08, 2022.

BBC. (2018b). Venezuela President Maduro survives' drone assassination attempt'. *BBC*. Available at: https://www.bbc.co.uk/news/world-latin-america-45073385. Accessed March 08, 2022.

Belsare, D., & Bhate, M. (2020). A review of NLP oriented automated test case generation framework in testing. *International Journal of Future Generation Communication and Networking, 13*(2), 14–16.

Business World. (2020). *What is Artificial Intelligence (AI) in Business?* Available at: www.businessworldit.com/ai/artificial-intelligence-in-business/. Accessed March 08, 2022.

Calder, S. (2019). *Gatwick drone disruption cost over £50M*. Available at: www.independent.co.uk/travel/news-and-advice/gatwick-drone-airport-cost-easyjet-runway-security-passenger-cancellation-a8739841.html. Accessed March 08, 2022.

CB Insights. (2017). *AI-driven facial recognition is coming and brings big ethics and privacy concerns*. Available at: www.cbinsights.com/research/facial-recognition-privacy-ai/. Accessed March 08, 2022.

CDT. (2014). *Mixed messages? The limits of automated social media content analysis*. Available at: https://cdt.org/wp-content/uploads/2017/11/Mixed-Messages-Paper.pdf

Chowdhury, G. G. (2003). Natural language processing. *Annual Review of Information Science and Technology, 37*(1), 51–89.

Cloudflare. (2021). *What is a DDoS attack?* Available at: www.cloudflare.com/en-gb/learning/ddos/what-is-a-ddos-attack/. Accessed March 08, 2022.

Conran, M. (2018). *The rise of artificial intelligence DDoS attacks*. Available at: www.networkworld.com/article/3289108/the-rise-of-artificial-intelligence-ddos-attacks.html. Accessed March 08, 2022.

Dar, P. (2019). *Popular machine learning applications and use cases in our daily life*. Available at: www.analyticsvidhya.com/blog/2019/07/ultimate-list-popular-machine-learning-use-cases/. Accessed March 08, 2022.

Devi, V. G., & Sairam, G. (2018). *Face recognition technology*.

Domingos, P. (2012). A few useful things to know about machine learning. *Communications of the ACM, 55*(10), 78–87.

Expert.ai. (2017). *What is machine learning? A definition*. Available at: https://www.expert.ai/blog/machine-learning-definition/. Accessed March 08, 2022.

Farley, B. W. A. C., & Clark, W. (1954). Simulation of self-organizing systems by digital computer. *Transactions of the IRE Professional Group on Information Theory, 4*(4), 76–84.

Feigenbaum, E. A., Barr, A., & Cohen, P. R. (Eds.). (1981). *The handbook of artificial intelligence*.

Felbo, B., Mislove, A., Søgaard, A., Rahwan, I., & Lehmann, S. (2017). Using millions of emoji occurrences to learn any-domain representations for detecting sentiment, emotion and sarcasm. *arXiv preprint arXiv:1708.00524*.

Felten, E. (2016). Preparing for the future of artificial intelligence. Washington DC*: The White House*, May, 3.

Gardner, K. (2019). *How AI is helping efficiency improve*. Available at: https:// towardsdatascience.com/how-ai-is-helping-efficiency-improve-98d0171a23e2. Accessed March 08, 2022.

Garg, S. (2020). *Face recognition using Artificial Intelligence*. Available at: www.geeksforgeeks.org/face-recognition-using-artificial-intelligence/. Accessed March 08, 2022.

GenieNetworks. (2019). *A new step towards network security: DDoS protection with machine learning and artificial intelligence*. Available at: www.genie-networks.com/wp-content/uploads/2019/06/WP_ML_DDoS_GN2019.pdf

Ghalleb, A. E. K., & Amara, N. E. B. (2020). A benchmark terrorist face recognition database. In *2020 International Conference on Cyberworlds (CW)* (pp. 285–288). IEEE.

Gillis, A. S. (2019). *Facial recognition*. Available at: https://searchenterpriseai.techtarget.com/definition/facial-recognition. Accessed March 09, 2022.

Glăvan, D., Răcuciu, C., Moinescu, R., & Antonie, N. F. (2019). DDoS detection and prevention based on artificial intelligence techniques. *Scientific Bulletin "Mircea cel Batran" Naval Academy, 22*(1), 1–11.

Gonzalez, P. T. (2013). *Dealing with variability factors and its applications to biometrics at a distance* = Tratamiento de factores de vaiabilidad y su aplicación en biometría a distancia (Doctoral chapter, Universidad Autónoma de Madrid).

Gorwa, R., Binns, R., & Katzenbach, C. (2020). Algorithmic content moderation: Technical and political challenges in the automation of platform governance. *Big Data and Society, 7*(1), 2053951719897945.

Gov.uk. (2018). *Safer prisons, safer streets*. Available at: https://www.gov.uk/government/speeches/safer-prisons-safer-streets. Accessed March 09, 2022.

Hill, J., Ford, W. R., & Farreras, I. G. (2015). Real conversations with artificial intelligence: A comparison between human–human online conversations and human–chatbot conversations. *Computers in Human Behavior, 49*, 245–250.

Ho, R. C. (2021). Chatbot for online customer service: Customer engagement in the era of artificial intelligence. In *Impact of globalization and advanced technologies on online Business models* (pp. 16–31). IGI Global.

Hovland, K. M., & Seetharaman, D. (2016). *Facebook backs down on censoring 'napalm girl' photo*. Available at: www.wsj.com/articles/norway-accuses-facebook-of-censorship-over-deleted-photo-of-napalm-girl-1473428032. Accessed March 09, 2022.

IBM Cloud Education. (2020). *Supervised learning*. Available at: www.ibm.com/cloud/learn/supervised-learning. Accessed March 09, 2022.

Järveläinen, E., & Rantanen, T. (2020). Incarcerated people's challenges for digital inclusion in Finnish prisons. *Nordic Journal of Criminology*, 1–20.

Jeong, S. S., & Seo, Y. S. (2019). Improving response capability of chatbot using twitter. *Journal of Ambient Intelligence and Humanized Computing*, 1–14.

Jordan, M. I., & Mitchell, T. M. (2015). Machine learning: Trends, perspectives, and prospects. *Science, 349*(6245), 255–260.

Kanowitz, S. (2021). *AI on the line: Monitoring prisoners' phone calls for criminal intent*. Available at: https://gcn.com/articles/2021/08/20/ai-prison-phone-conversations.aspx. Accessed March 09, 2022.

Karanicolas, M. (2021, June). Personal communication. Retrieved from https://galley.cjr.org/public/conversations/-MblGkCdmRJrJfixYLr-

Klare, B. F., Burge, M. J., Klontz, J. C., Bruegge, R. W. V., & Jain, A. K. (2012). Face recognition performance: Role of demographic information. *IEEE Transactions on Information Forensics and Security, 7*(6), 1789–1801.

Kok, J. N., Boers, E. J., Kosters, W. A., Van der Putten, P., & Poel, M. (2009). Artificial intelligence: Definition, trends, techniques, and cases. *Artificial Intelligence, 1*, 270–299.

Kvale, K., Sell, O. A., Hodnebrog, S., & Følstad, A. (2019). Improving conversations: Lessons learnt from manual analysis of chatbot dialogues. In *International workshop on chatbot research and design* (pp. 187–200). Springer.

Lavion, D., Rivera, K., & Elliott, S. (2018). Pulling fraud out of the shadows: Global Economic Crime and Fraud Survey 2018. *PwC Global Economic Crime and Fraud Survey Report*. Available at: https://pwc.com/.../global-economic-crime-and-fraud-survey-2018-summary-inf. Accessed March 08, 2022.

Lawton, G., & Wigmore, I. (2021). *AI ethics (AI code of ethics)*. Available at: https://whatis.techtarget.com/definition/AI-code-of-ethics. Accessed March 08, 2022.

Lester, J., Branting, K., & Mott, B. (2004). Conversational agents. In *The Practical Handbook of Internet Computing* (pp. 220–240).

Li, J. J., Bonn, M. A., & Ye, B. H. (2019). Hotel employee's artificial intelligence and robotics awareness and its impact on turnover intention: The moderating roles of perceived organizational support and competitive psychological climate. *Tourism Management, 73*, 172–181.

Llansó, E., Van Hoboken, J., Leerssen, P., & Harambam, J. (2020). *Artificial intelligence, content moderation, and freedom of expression.*

Locascio, N. (2018). *Black-box attacks on perceptual image hashes with GANs*. Available at: https://towardsdatascience.com/black-box-attacks-on-perceptual-image-hashes-with-gans-cc1be11f277. Accessed March 08, 2022.

Longoni, C., Bonezzi, A., & Morewedge, C. K. (2019). Resistance to medical artificial intelligence. *Journal of Consumer Research, 46*(4), 629–650.

Mapua, J. (2019). *Can anyone own a gun?* Enslow Publishing, LLC.

Merriam-Webster. (n.d.). *Phishing*. Available at: www.merriam-webster.com/dictionary/phishing. Accessed March 08, 2022.

McCarthy, J. (1960). Recursive functions of symbolic expressions and their computation by machine, part I. *Communications of the ACM, 3*(4), 184–195.

McKendrick, K. (2019). *Artificial intelligence prediction and counterterrorism*. Available at: www.chathamhouse.org/sites/default/files/2019-08-07-AICounterterrorism.pdf

McCulloch, W. S., & Pitts, W. (1943). A logical calculus of the ideas immanent in nervous activity. *The Bulletin of Mathematical Biophysics, 5*(4), 115–133.

Microsoft. (2021). *Cortana in Microsoft 365*. Retrieved July 20, 2021, from Cortana in Microsoft 365 – Microsoft 365 admin | Microsoft Docs.

Newell, A., & Shaw, J. C. (1957). Programming the logic theory machine. In *Papers presented at the February 26–28, 1957, western joint computer conference: Techniques for reliability* (pp. 230–240).

Newell, A., Shaw, J. C., & Simon, H. A. (1958). Elements of a theory of human problem solving. *Psychological Review, 65*(3), 151.

NI Business Info. (n.d.). *Artificial intelligence in business*. Available at: www.nibusinessinfo.co.uk/content/business-benefits-artificial-intelligence. Accessed March 08, 2022.

Noorbakhsh-Sabet, N., Zand, R., Zhang, Y., & Abedi, V. (2019). Artificial intelligence transforms the future of health care. *The American Journal of Medicine, 132*(7), 795–801.

Pledger, T. (2021). *The role of drones in future terrorist attacks* (pp. 1–16). Association of the United States army.

Press, G. (2019). *AI stats news: 86% of consumers prefer humans to chatbots*. Available at: www.forbes.com/sites/gilpress/2019/10/02/ai-stats-news-86-of-consumers-prefer-to-interact-with-a-human-agent-rather-than-a-chatbot/?sh=58a3ce162d3b. Accessed March 08, 2022.

Privacy International and ARTICLE 19. (2018). *Privacy and freedom of expression in the age of artificial intelligence*. Available at: www.article19.org/wp-content/uploads/2018/04/Privacy-and-Freedom-of-Expression-In-the-Age-of-Artificial-Intelligence-1.pdf. Accessed March 08, 2022.

Puolakka, P. (2020). *Smart Prisons in Finland (2021)*. Available at: www.europris.org/news/smart-prisons-in-finland-2021/. Accessed March 09, 2022.

Rantanen, T., Järveläinen, E., & Leppälahti, T. (2021). Prisoners as users of digital health care and social welfare services: A finnish attitude survey. *International Journal of Environmental Research and Public Health, 18*(11), 5528.

Rhue, L. (2018). *Racial influence on automated perceptions of emotions*. Available at SSRN 3281765.

SAS. (n.d.). *Natural language processing (NLP). What it is and why it matters*. Available at: www.sas.com/en_us/insights/analytics/what-is-natural-language-processing-nlp.html. Accessed March 08, 2022.

Segers, J. (2014). *Perceptual image hashes*. Available at: https://jenssegers.com/perceptual-image-hashes. Accessed March 08, 2022.

Swindler, D. (2018). *Face recognition is the new way to prevent crime*.

Tellez, E. S., Miranda-Jiménez, S., Graff, M., Moctezuma, D., Suárez, R. R., & Siordia, O. S. (2017). A simple approach to multilingual polarity classification in twitter. *Pattern Recognition Letters, 94*, 68–74.

Thomas, N. T. (2016). An e-business chatbot using AIML and LSA. In *2016 International Conference on Advances in Computing, Communications and Informatics (ICACCI)* (pp. 2740–2742). IEEE.

Tlight. (2020). *What are drones and what are they used for?* Retrieved June 26, 2021, from What are drones? Drone Uses | FAQs | RAWview.

Topham, G. (2019). Gatwick drone disruption cost airport just £1.4m. *The Guardian*. Available at: www.theguardian.com/uk-news/2019/jun/18/gatwick-drone-disruption-cost-airport-just-14m. Accessed March 08, 2022.

Turing, A. M. (1948). Intelligent machinery, a heretical theory. *The Turing test: Verbal behavior as the hallmark of intelligence, 105*.

Tyagi, N. (2020a). *6 major branches of artificial intelligence (AI)*. Available at: https://www.analyticssteps.com/blogs/6-major-branches-artificial-intelligence-ai. Accessed March 08, 2022.

Tyagi, N. (2020b). *7 popular applications of machine learning in daily life*. Available at: https://www.analyticssteps.com/blogs/7-popular-applications-machine-learning-daily-life. Accessed March 08, 2022.

van der Veer, R. (2019). *Terrorism in the age of technology*. Clingendael Institute.

Vigna, G. (n.d.). *How AI will help in the fight against malware*. Available at: https://techbeacon.com/security/how-ai-will-help-fight-against-malware. Accessed March 08, 2022.

Veilleux-Lepage, Y. (2021). *CTRL, HATE, PRINT: Terrorists and the appeal of 3D-printed weapons*. Available at: https://icct.nl/publication/ctrl-hate-print-terrorists-and-the-appeal-of-3d-printed-weapons/. Accessed March 08, 2022.

Webster, R. (2017). *British prison shields itself against drug drones*. Available at: https://www.russellwebster.com/british-prison-shields-itself-against-drug-drones/. Accessed March 08, 2022.

Weiss, D. C. (2019). *Prisons and jails use artificial intelligence to monitor inmate phone calls*. Available at: https://www.abajournal.com/news/article/prisons-and-jails-use-artificial-intelligence-to-monitor-inmate-phone-calls. Accessed March 09, 2022.

Countering Terrorism: Digital Policing of Open Source Intelligence and Social Media Using Artificial Intelligence

Sarah Klingberg

Abstract This chapter critically discusses the use of social media and artificial intelligence in digital policing to counter terrorism. Key concepts are defined followed by a critical analysis of the ethical, legal, technological and organizational challenges. Additionally, a number of recommendations is included, such as a transparent and clear framework of policing social media nationally and internationally, screening for biased algorithms and more transparency of processes within artificial intelligence as well as restructuring of and more funding for the police to improve digital investigations. Moreover, the importance of preventing human rights violations has been discussed by evaluating the restriction of freedom of speech on social media platforms within legal and ethical contexts, suggesting a more open approach of redirecting users at risk of radicalization to verified sources.

Keywords Artificial intelligence · Digital policing · Social media · Cybercrime · Online radicalization · Digital terrorism · Human rights · Counterterrorism · OSINT · SOCMINT

1 Introduction

There is a high need for countering terrorism digitally (United Nations Interregional Crime and Justice Research Institute (UNICRI) & United Nations Counter-Terrorism Centre (UNCCT), 2021). Terrorism being a significant threat online was demonstrated by the 2020 Referral Action day where Europol and 17 member states identified and removed 1906 URLs which lead to 180 platforms containing terrorist content. Furthermore, Facebook, a popular social media platform, claimed to have deleted more than 26 million pieces of content from terrorist associations such as

S. Klingberg (✉)
School of Social Sciences, Department of Criminology, Sociology and Social Policy, Swansea University, Swansea, UK
e-mail: klingbergsarah@web.de
URL: http://www.swansea.ac.uk

R. Montasari (ed.), *Artificial Intelligence and National Security*,
https://doi.org/10.1007/978-3-031-06709-9_6

the Islamic State of Iraq and the Levant (ISIL) within 2 years (UNICRI & UNCCT, 2021). This emphasizes social media as a tool used by terrorist organizations to distribute their ideologies and recruit members. According to Hamidi, the former Home Affairs Minister of Malaysia, around 17 per cent of ISIL recruitment was performed using social media (Signpeng & Tapsell, 2020, as cited in UNICRI & UNCCT, 2021). The following report aims to display challenges for policing terrorism on the internet, focussing on social media intelligence (SOCMINT) as part of open source intelligence (OSINT) and demonstrating recommendations for future actions to enhance countering terrorism online. First, some background information of key terms will be provided before addressing legal and ethical, technological and organizational challenges and recommendations. This report is adding to the existing debate on using artificial intelligence for digital policing and applying it to social media use on countering terrorism while protecting human rights.

2 Background

2.1 Cybercrime

Cybercrime is highly relevant, carrying an estimated annual cost of 27 billion pounds alone for the UK economy in 2011 (Detica & Office of Cyber Security and Information Assurance, 2011). The fast-changing technology and myriad private and public organizations entangled in regulating cyberspace impede a fixed definition (Williams & Wall, 2013). Maras (2015) as cited in Vincze (2016) defines cybercrime as using the internet, computers and other similar technologies to commit a crime. Thereby, computers can be used to commit a crime or be the victim. Hereby, included crimes are child pornography, homicide, hacking and theft. Williams (2010), as cited in Williams and Wall (2013), adds hacktivism, hate crime, cyberbullying and cyberterrorism to that list. Moreover, cybercrime poses a major threat, facilitating criminal acts by being faster without the need for physical presence (Vincze, 2016), and globally, across cultures and time zones (Williams & Wall, 2013).

2.2 Digital Policing

Digital policing is based on the assumption that social media is a virtual public space prone to include criminalization. By countering criminal actions online, existing laws and protocols are sought to be implemented in online investigations (Trottier, 2015).

2.3 Open Source Intelligence

Open Source Intelligence (OSINT) is a method of collecting, analysing and interpreting lawfully publicly available information on the internet, including media as radio and newspapers, social media, statistics and academic publications (Richelson, 2016, as cited in UNICRI & UNCCT, 2021). Furthermore, different searches are available such as scanning of images, texts from websites and social media and maps (García, 2021). Moreover, there are OSINT tools that facilitate navigating the data (UNICRI & UNCCT, 2021). Although the information is publicly available, certain subscriptions or accounts to access newspapers or social media might be needed (OSINT, 2020). Thereby, OSINT is easily accessible, cheap, low risk and effective as described by the corporate intelligence consultant Colquhoun (OSINT, 2020).

2.4 Social Media Intelligence

Social Media Intelligence (SOCMINT) SOCMINT is a subcategory of OSINT concerned with data collection on social media, analysing conversations and combining data to useful trends (UNICRI & UNCCT, 2021). Social media platforms (SMPs) like Spotify, Facebook and Instagram are used to connect people by sharing texts, pictures and videos (Murray, 2019a). According to Bowling et al. (2019), social media changed people's interactions, posing the advantage of independence from space and time, but also a disadvantage like a broad audience to terrorist organizations (Murray, 2019a). Terrorist groups use social media to spread ideologies, recruit sponsors and members or plan and coordinate attacks (UNICRI & UNCCT, 2021). The popularity of these platforms can be demonstrated by the 1,470,000,000 photo uploads per day on Instagram in July 2018 (Murray, 2019a). Moreover, social media poses other risks for victimization such as cyberstalking and bullying, especially for minors, who are more insecure and susceptible to manipulation (Murray, 2019a).

2.5 Terrorism and Radicalization

Both terms have no unifying definition. First, terrorism can be described as a serious violent act against the general public (Scheinin, 2017) performed by non-state highly organized groups to reach a clear long-term ideological aim, as well as spontaneous, little organized individuals. Hereby, the aim is mostly to attack the existing governing belief system or regime by terrorizing or polarizing the public (Innes & Levi, 2017). Since terrorism violates several human rights such as liberty and physical integrity, it has been described as the antithesis to human rights (Scheinin, 2017). Second, radicalization is the path to becoming an extremist or terrorist. It can be viewed as a two-stage process: first, a vulnerable individual, especially a child or teenager, begins to hold extremist views and second, starts to act on them (Chisholm & Coulter, 2017).

3 Digital Policing Methods of Countering Terrorism

In digital policing, SOCMINT and OSINT are usually treated as one (Dencik et al., 2018). By using OSINT, it is aimed to collect information about recent views (Eijkman & Weggemans, 2013), members, possible actions and key figures (UNICRI & UNCCT, 2021). Therefore, social network analysis (SNA), a framework to depict a network structure using social media sources, is used (Musotto, 2020, as cited in UNICRI & UNCCT, 2021). Currently, social media monitoring is used at protests, tracking the groups' movement, creating personal profiles of key figures and monitoring potential aggression levels towards the police. The data is collected by filtering large datasets using keywords. Afterwards, findings are reviewed (Dencik et al., 2018). García (2021) categorizes the 'process of private investigation' into six phases as described below (Fig. 1).

Requirements refer to the information that needs to be collected while Information Sources relates to the sources of information to be used as well as the costs involved. Acquisition of Information involves collecting information, and Processing addresses the extraction of useful data from collected information.

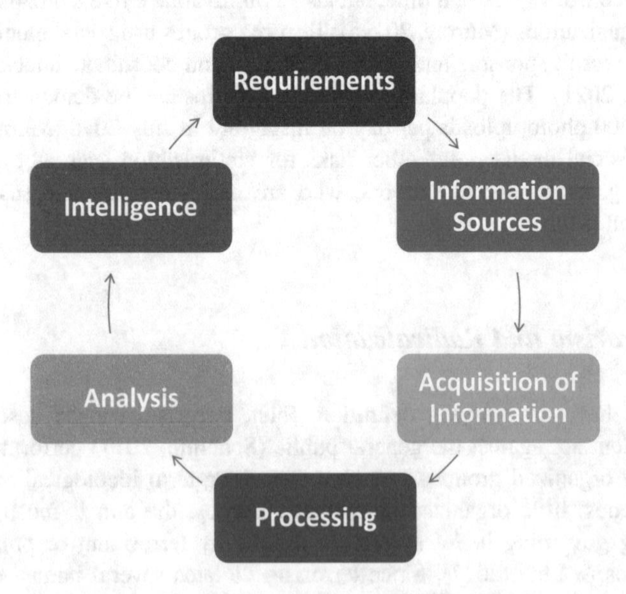

Fig. 1 Process of private investigation based on García's (2021) work

Analysis gives meaning to the data, and Intelligence determines the actions that will need to be taken based on the Analysis phase (García's, 2021). SOCMINT also includes 'geotagging', providing a location to an individuals' online post, but it is difficult to implement since less than 2 per cent of tweets on Twitter are geotagged (Dencik et al., 2018). Overall, a summary of sources most successfully provides an accurate understanding of targets (Trottier, 2015). One promising tool to aid digital policing is artificial intelligence (AI). AI is used to process large amounts of data and discovers patterns and links within (UNICRI & UNCCT, 2021). It rapidly performs repetitive and cognitively demanding tasks such as visual perception, speech recognition, language translation, decision making and problem-solving, which else require human interaction (UNICRI & UNCCT, 2021). Furthermore, it calculates likely outcomes and detects suspicious financial transactions (UNICRI & UNCCT, 2021). The AI used today is called narrow AI since it is limited to a certain task, having problems adapting to a changing environment. Thus, AI engages in tasks it is programmed for like content matching technology, which identifies identical content in real-time. It is used by service providers like Google, to tackle terrorist imagery using PhotoDNA (Facebook, 2016, as cited in UNICRI & UNCCT, 2021). Different strategies using AI to counter terrorism online are outlined below (UNICRI & UNCCT, 2021).

AI manages big data analysis and performs predictive analytics. Especially after terror attacks like 9/11, the demand for using OSINT for predictive policing increased, aiming to become more proactive towards terrorist attacks (Dencik et al., 2018). Exemplifying, AI use in predictive policing was applied in a study by Spezzano et al. (2013), as cited in UNICRI and UNCCT (2021). They used behavioural models and data about the Lashkar-e-Taiba group in a Rule Learning Algorithm to produce policy suggestions to reduce future attacks of this terrorist group. Prior, this group has been known, for example, the 2008 Mumbai attacks (Spezzano et al., 2013, as cited in UNICRI & UNCCT, 2021). Moreover, AI can be used to identify individuals that are becoming radicalized, as shown by the European Union-funded Real-time Early Detection and Alert System for Online Terrorist Content (RED-Alert). By using, for instance, SNA, early stages of radicalization can be assessed. However, it is only in the test phase yet. After having flagged individuals being subject of radicalization, the terrorist views are countered by directing the individual to deradicalization content (Barata, 2021, as cited in UNICRI & UNCCT, 2021). As used by Google in 2016 implementing Moonshot's 'redirect method' (UNICRI & UNCCT, 2021). If people have been looking at certain pre-defined contents, de-radicalization advertisements and videos have been suggested. However, this raises serious ethical and legal issues discussed in Sect. 4.

Furthermore, AI automatically monitors and deletes terrorist content. There are two ways of deleting terrorist propaganda: de-platforming, which means to take down someone's content and block their access to an online forum, and shadowbanning, deleting posts and limiting the accessibility of content. Both approaches are highly difficult to realize (UNICRI & UNCCT, 2021). Lacking a unifying definition of terrorism aggravates distinguishing objectively between political and hate speech (UNICRI & UNCCT, 2021). One example of deleting terrorist content occurred when perpetrators live-streamed the Christchurch terrorist attack on Facebook in 2019. Even though the company managed to take down the original content 12 minutes after the recording ended, in the following 24 hours, 1.5 million international attempts to upload copies of the video were countered (Facebook, 2019, as cited in UNICRI & UNCCT, 2021). Additionally, detecting mis- and disinformation spread by terrorists using bots, to impair the public's trust in the government and distribute terrorist ideology can be counteracted by AI in identifying those bots. This will be used by a British intelligence and security organization called Government Communications Headquarters (GCHQ) who declared to use AI tools to expose fake accounts. Additionally, it is intended to use automatic fact-checking technology by validating information with trusted sources (Government Communications Headquarter (GCHQ, 2021).

4 Challenges and Recommendations

The previous sections provided insight in constructs and tools to conquer terrorism in cyberspace. The following will discuss challenges linked to digital policing using social media and AI, including recommendations to overcome these.

4.1 Ethical and Legal Challenges

First of all, surveillance by digital sources needs to be ethical and legal. By using OSINT, especially SOCMINT, police engages in mass surveillance, which needs to be regulated, as it is criticized to be similar to spying (O'Neill, 2012, as cited in Eijkman & Weggemans, 2013). However, secret surveillance to prevent radicalization in democratic states is legal, exemplifying the Counter-Terrorism and Security Act 2015 in the UK (Chisholm & Coulter, 2017). As expressed in UK Art 8(2) in ECHR, surveillance is used to protect the states' rights and freedom and maintain public safety, but only to an extent where no state control, as in former East Germany, is established (Murray, 2019b). Thus, guidelines on how to use OSINT as a legal and ethical measure of surveillance need to be established (Brayne, 2017, as cited in Bowling et al., 2019). In the UK, for instance, it is allowed to make use of public accounts on websites, but not to intercept communication without a warrant as stated in the Investigatory Power ACT 2016 (Murray, 2019b). Hence, when national security is endangered, content hidden by privacy settings may be reclassified as public when shared privately with friends (Kelly, 2012, as cited in Trottier, 2015). The downside, it could severely affect non-criminal individuals to avoid using social media for mental health or religious reasons, fearing stigmatization if the information becomes known publicly. Furthermore, it increases distrust in police and government (Trottier, 2015). To avoid this, it is important to make clear, transparent guidelines on using SOCMINT. This might decrease the fear of surveillance and increase a feeling of safety and a democratic understanding of national security. Second, by using SOCMINT to counter terrorism, human rights must not be violated, thereby, the most affected rights often depicted are the right to freedom of thought and expression and the right to non-discrimination (UNICRI & UNCCT, 2021). However, algorithms of AI mirror society, since society is biased and discriminatory AI algorithms may produce discriminatory outcomes, leading to severe consequences. One example has been Amazon's AI automatic hiring tool in 2018, which was gender-biased due to the mere fact that there was a majority of male workers in the company. So, the system taught itself to prefer male applicants (Dastin, 2018, as cited in UNICRI & UNCCT, 2021). This further demonstrated that AI did not only apply the bias but also amplified it. In this case, the algorithm was biased unintentionally, but thinking about AI implementation in authoritarian governments using consciously biased technology to eliminate opponents demonstrates a great danger. Thus, AI algorithms need to be democratic by nature, being screened for underlying biases.

Third, regulating freedom of speech and expression is highly demanding. Social media provide a broad audience for terrorists. Therefore, the UK restricts this kind of opinion spreading by the Public Order Act 1986, which forbids racially offensive speech and owning such content with the intent of publication (Murray, 2019a). It was then extended by the Racial and Religious Hatred Act 2006, countering speech that results in religious hatred (Murray, 2019a). On an international level, the Human Rights Council (2012), as cited in UNICRI and UNCCT (2021), declares that human rights apply the same online and offline. Regarding freedom, an important distinction needs to be made: internal freedom, which is freedom of thought that no government is allowed to limit, and external freedom or freedom of expression that manifest those thoughts and can be restricted since it limits others' freedom and well-being (UNICRI & UNCCT, 2021). A quite interesting legal case was started in May 2000 when the League Against Racism and Antisemitism (LICRA) and the Union of French Jewish Students (UEJF) legally fought Yahoo! Inc. and Yahoo! France, which offered links to buy Hitler's 'Mein Kampf' online and this violated French law. After discussions, Yahoo! agreed to warn French users when accessing illegal content (Murray, 2019a). However, this case demonstrates a huge challenge towards today's restriction of content on OSINT and SOCMINT since each country follows a different legal framework; big companies such as Facebook are constantly challenged to keep freedom of speech and reduce hate, discrimination and harm towards their users. Those organizations possess an enormous power to alter individuals' mindset. Taking into account the AI method of redirecting individuals towards de-radicalization is a powerful tool with high risk when used for undemocratic, authoritarian purposes. It raises the question of whether it is ethical to invade an individual's self-determination and opinion-forming. Should global players incorporate that much power on a global audience?

Moreover, algorithms recommending systems, used for instance by YouTube, include the risk of becoming more radicalized by creating so-called filter bubbles which direct users to content similar to what they engage in (Reed et al., 2019, as cited in Schroeter, 2020). By recommending users a wider variety of sources and topics, broadening their mindset is ensured, the probability of radicalization and filter bubbles decreased and it provides an ethical alternative to altering mindset. Furthermore, by using de-platforming and shadowbanning, it is vital to ground those decisions on specific guidelines, since no universal definition for terrorism exists, distinguishing between hate and political speech is highly difficult, often depending on cultural and political obstacles (Murray, 2019a). Although being far from an international definition, lots of recommendations to improve the existing fragmented definition exist, such as focussing more on the (violent) methods used by terrorists (Scheinin, 2017).

4.2 Technological Challenges

First of all, AI is currently the focus of an extensive debate among technologists, ethicists and policymakers worldwide (UNICRI & UNCCT, 2021). At the moment, AI is used rather narrowly, which means it is unable to adapt to a changing world and is specifically programmed for a specific task. Furthermore, it lacks transparency. Hence, unintended neural connections can be created, exemplifying by feeding a machine pictures of trains to detect trains on unknown graphics the AI might focus on train tracks (Thesing et al., 2019, as cited in UNICRI & UNCCT, 2021). These minor deviations can have serious effects when collecting wrong data on terrorism. In future, AI processing must be traceable to ensure ethical, democratic outcomes.

Second, the complexity of the content challenges AI programming. Focussing on language, there are several issues attached, for instance, different languages, dialects and even invented languages by terrorism groups that make it difficult to establish an algorithm for AI (UNICRI & UNCCT, 2021). Moreover, it is also known that Twitter's algorithm is not able to distinguish between offensive language, sarcasm and friendly fighting (Butler & Parrella, 2021, as cited in UNICRI & UNCCT, 2021). This may lead to serious consequences as shown by Paul Chambers who publicly posted a message threatening to bomb an airport. What started as a joke, led to his arrest, conviction, and afterwards, he was set free on the third appeal – he has lost his job (Mitchell, 2010, as cited in Trottier, 2015). To understand language processing in detail, more funding for research in human language processing and recognition by machines is needed.

4.3 Organizational Challenges

In this section, challenges within police, but also between police and private companies in countering terrorism online, will be discussed.

First, although using OSINT is cheaper in collecting information compared to classified sources, analysing data is costly, especially by adding staff wages and training costs (Trottier, 2015). In this case, AI can be highly sufficient, reducing staff for repetitive tasks. But still, one needs to have access to those kinds of technologies first. Moreover, a lot of tools are limited in use, being invented for marketing purposes, but have been adapted by the police for profiling potential terrorists (Dencik et al., 2018). Thus, it needs more funding for proper tools used for policing. Furthermore, human resources still need to act on AI-collected data, but these are decreasing in number (Trottier, 2015) and need to be trained, which is difficult due to budget cuts (Vincze, 2016). Training is especially important since the majority of police officers did not grow up with social media, being 40-plus, white males (Dencik et al., 2018). Thus, a higher budget for police staff, training and research is needed. Additionally, recruiting processes in some police hierarchies do not represent digital policing and there is no proper job application for digital investigators (Trottier, 2015), this requires a restructuring of police. Second, there must be an improvement between the work of police and private companies. The police do rely on companies like Google to adapt to investigation – if content is deleted, evidence crucial for conviction might be lost (UNICRI & UNCCT, 2021). The power of the private sector is significantly increasing due to their financial resources. It acts as a gatekeeper providing access to data needed to develop AI, data collection with AI. Moreover, private companies develop AI technologies used by the police, meaning the state heavily relies on private organizations to counter terrorism in cyberspace, showing a need for cooperation (UNICRI & UNCCT, 2021).

5 Discussion

First of all, digital policing methods, challenges in policing social media using AI without impeding human rights, as well as technological implications and structural challenges within the police have been outlined and discussed. Thereby it was found that surveillance of the general public in cyberspace is legal within certain boundaries for protecting national security, but incorporates the risk that citizens lose trust in police and government and avoid services to improve, for instance, mental health fearing stigmatization. Therefore, it has been recommended to establish clear guidelines on data use in SOCMINT that are transparent and thus, accessible for citizens. Second, AI algorithms based on biased data from society amplified those biases as shown with Amazon's hiring tool (Dastin, 2018, as cited in UNICRI & UNCCT, 2021), leading to undemocratic, discriminatory outcomes. Hence, establishing unbiased algorithms and screening for underlying biases is vital. Third, the legal case of Yahoo! in France (Murray, 2019a) demonstrates the challenges of restricting and deleting social media content without violating freedom of speech and expression. On the one hand, it has been discussed how redirecting users to content can avoid radicalization and filter bubbles, but also give too much power to global companies in determining how an individual's mindset should be altered. Thus, directing users to a wide variety of verified sources to extend their knowledge and provide self-determination in opinion-forming is emphasized. On the other hand, the issues in determining the difference between political and hate speech have been addressed, by discussing the lack of a unifying definition. It is suggested to focus on methods used by terrorists rather than their aim. Fourth, the missing transparency of AI processing has been criticized and demanded to be clearer to check for biases and unwanted machine learning. Fifth, the challenges AI faces by interpreting different language parameters have been outlined and suggestions for more funding and research were made. Sixth, the lack of financial resources of the police and a conservative structure within the organization has been depicted. To improve digital policing, a higher budget for research and development on suitable AI tools and more staff and training, as well as an innovative restructuring of the police that incorporates digital investigators has been suggested. Finally, the link between police or public organization and the private sector was displayed and the importance of cooperation and good relationships was emphasized. Moreover, this work is limited due to the present status of technology research, providing ideas for AI usage, which will need some time until they can be fully applied and evaluated in real-life-investigations. Additionally, the fragmented landscape on laws that can be adapted towards using SOCMINT to counter terrorism and protecting human rights leads to difficulties and controversies in this discussion.

6 Conclusion

To start with, the provided information has shown how social media can be used to counter terrorism. Thereby, the use of AI was introduced and heavily discussed towards human rights and technological implications. Further challenges that have been discussed regarding legal grounds for surveillance, a lack of definition for terrorism and challenges police organizations face. Recommendations to resolve these issues have been provided. This report has added an interesting discussion on SOCMINT use to counter terrorism using AI, while protecting human rights to the existing literature. Future research is needed, especially, to assess AI reliability when it can be fully implemented to real-life use. Further research topics have been addressed, such as language processing and recognition in machines and their intersection with human language production. Moreover, local and global guidelines on AI and SOCMINT use, as well as an international approach on restricting freedom of speech to counter terrorism or other forms of human right violations, have to be investigated and established. To conclude, it is shown that SOCMINT, in combination with AI, has an enormous value to counter terrorism and to ensure proper use, organizational, technological, legal and ethical challenges must be resolved.

References

Bowling, B., Reiner, R., & Sheptycki, J. (2019). 10. Police and media. In B. Bowling, R. Reiner, & J. Sheptycki (Eds.), *The politics of the police* (pp. 207–226). Oxford University Press. https://doi.org/10.1093/he/9780198769255.003.0010

Chisholm, T., & Coulter, A. (2017). *Safeguarding and radicalisation*. Department for Education. https://www.gov.uk/government/publications/safeguarding-and-radicalisation

Dencik, L., Hintz, A., & Carey, Z. (2018). Prediction, pre-emption and limits to dissent: Social media and big data uses for policing protests in the United Kingdom. *New Media & Society, 20*(4), 1433–1450. https://doi.org/10.1177/1461444817697722

Detica & Office of Cyber Security and Information Assurance. (2011). *The cost of cyber-crime*. Cabinet Office and National security and intelligence. https://www.gov.uk/government/publications/the-cost-of-cyber-crime-joint-government-and-industry-report

Eijkman, Q., & Weggemans, D. (2013). Open source intelligence and privacy dilemmas: Is it time to reassess state accountability? *Security and Human Rights, 23*(4), 285–296. https://doi.org/10.1163/18750230-99900033

García, F. J. C. (2021). Private investigation and Open Source INTelligence (OSINT). In *Cybersecurity threats with new perspectives*. IntechOpen. https://doi.org/10.5772/intechopen.95857

Government Communications Headquarters (GCHQ). (2021). *Pioneering a new national security – The ethics of artificial intelligence*. GCHQ. https://www.gchq.gov.uk/files/GCHQAIPaper.pdf

Innes, M., & Levi, M. (2017). 20. Making and managing terrorism and counter-terrorism: The view from criminology. In M. Innes & M. Levi (Eds.), *The Oxford handbook of criminology*. Oxford University Press. https://doi.org/10.1093/he/9780198719441.003.0021

Murray, A. (2019a). 5. Cyber-speech. In A. Murray (Ed.), *Information technology law* (pp. 87–124). Oxford University Press. https://doi.org/10.1093/he/9780198804727.003.0005

Murray, A. (2019b). 25. State surveillance and data retention. In A. Murray (Ed.), *Information technology law* (pp. 639–676). Oxford University Press. https://doi.org/10.1093/he/9780198804727.003.0025

OSINT: What is open source intelligence and how is it used? (2020, November 19). The Daily Swig | Cybersecurity News and Views. https://portswigger.net/daily-swig/osint-what-is-open-source-intelligence-and-how-is-it-used

Scheinin, M. (2017). 29. Terrorism. In M. Scheinin (Ed.), *International human rights law*. Oxford University Press. https://doi.org/10.1093/he/9780198767237.003.0029

Schroeter, M. (2020). *Artificial intelligence and countering violent extremism: A primer*. Global Network on Extremism and Technology (GNET). https://gnet-research.org/2020/09/28/artificial-intelligence-and-countering-violent-extremism-a-primer/

Trottier, D. (2015). Open source intelligence, social media and law enforcement: Visions, constraints and critiques. *European Journal of Cultural Studies, 18*(4–5), 530–547. https://doi.org/10.1177/1367549415577396

United Nation Interregional Crime and Justice Research Institute (UNICRI), & United Nations Counter-Terrorism Centre (UNCCT). (2021). *Countering terrorism online with artificial intelligence*. United Nations Office of Counter-Terrorism. https://www.un.org/counterterrorism/publications

Vincze, E. A. (2016). Challenges in digital forensics. *Police Practice and Research, 17*(2), 183–194. https://doi.org/10.1080/15614263.2015.1128163

Williams, M., & Wall, D. (2013). 12. Cybercrime. In C. Hale, K. Hayward, A. Wahidin, & E. Wincup (Eds.), *Criminology* (pp. 247–266). Oxford University Press. https://doi.org/10.1093/he/9780199691296.003.0012

Cyber Threat Prediction and Modelling

Jim Seaman

Abstract Threat prediction and modelling is an extremely important part of risk management and should be a focus for any organization. Effective practices in threat prediction and modelling enable a company to understand the traditional threats (e.g. the tactics and techniques that a threat actor may use against them), and to understand how any non-traditional threats might affect their valued business operations/processes.

This chapter focuses on how various threat (information) resources can be collected and used to help an organization create usable and actionable intelligence that can then be used to create a proactive defensive model. This defensive model can then be used to enhance a business' strategy and planning to help reduce the risks and to help provide the C-Suite personnel with added assurance, and to help show the return on investment provided by the defence strategy.

Unlike forecasting the weather (Meteorology), often, risk management is not seen as being an exact science. However, much like meteorology, risk assessments need to predict the potential impact that any identified threats may predict. This chapter shows how threat prediction and modelling should be given the same respect as predictive modelling gets within meteorology. Additionally, just as weather predictions are visualized in Meteorology (e.g. Weather Maps), this chapter shows the benefits for visualizing threat predictions and models.

Keywords Threat modelling · Forecasting · Prediction · Cybersecurity · Business · Risk management · Threat actors

J. Seaman (✉)
IS Centurion Consultancy Ltd, Castleford, UK
e-mail: contact@iscenturion.com
URL: https://www.iscenturion.com

R. Montasari (ed.), *Artificial Intelligence and National Security*,
https://doi.org/10.1007/978-3-031-06709-9_7

1 Introduction

Cyber threat prediction and modelling has never been more important with Artificial Intelligence (AI) and National Security. Data has become the 'life-blood' of the digital business, which requires longevity of their critical assets (vital organs).

The contamination or loss of this 'life-blood' can have significant impact on an organization's ability to support continual operations, which results in significant impact on the affected businesses.

Consequently, every business should be implementing a robust cyber threat prediction and modelling programme into their defensive strategies.

Why (I hear you ask)?

Think of this as being like trying to forecast the weather. In meteorology, they look at the available information and use this to try and forecast the future weather conditions (Shetty, 2018), e.g.

- Weather stations
- Satellites
- Sea buoys
- Commercial airliners
- Ships

All this information is used to create actionable intelligence to help provide a forecast of the forthcoming weather. However, we all know that the predictive weather modelling is not always correct and sometimes the meteorologists do have inaccurate weather predictions. Yet, most people will still value and take note of the weather forecasts, e.g.

- Taking a warm jacket with you, if the weather forecast is predicted colder weather conditions.
- Taking a raincoat or umbrella with you, if there is rain forecasted.

The same applies for threat modelling, in support of cybersecurity, defensive and risk management strategies. Businesses need to take advantage of a myriad of information (both open source and paid for) to help them to create actionable intelligence so that they are better able to align their defences to the forecasted threats.

2 Business Importance of Cyber Threat Prediction and Modelling

Effective risk management is extremely dependent on having an operational Cyber threat prediction and modelling practice, with risk management being reliant on the evaluation of the following:

1. Business assets'/operations' perceived value
2. The perceived threats to the valued business assets/operations
3. The perceived vulnerabilities of the valued business assets/operations, which could be exploited by the perceived threats
4. The perceived impact to the business, should a threat exploit the vulnerabilities of the valued business assets/operations

Imagine trying to safely navigate your way across a road/highway without looking out for the threats, understanding your weaknesses or having an appreciation for the potential impacts. As children, we are taught the importance of 'Stop, Look and Listen' to help us to safely navigate our way across the road/highway, which equates to threat prediction and modelling.

Using **'Stop, Look and Listen'** our body's senses are assessing the threats so that we can make that informed decision as to whether we can safely get across the road/highway, without getting hit by a vehicle, e.g.

- **Valued asset**
  - The child
- **Threats**
  - The speed of the vehicles
  - The size of the vehicles
- **Vulnerabilities**
  - The landscape
  - Any injuries that affect the child's capability to cross the road
  - Any obstructions
  - The lighting conditions
  - The weather conditions
- **Impact**
  - Significant injury
  - Death

Now, given the importance of **'Stop, Look and Listen'** why wouldn't you do the same for Artificial Intelligence (AI) and National Security? Before 'stepping out into the road' of making a business decision, wouldn't you want to **'Stop, Look and Listen'** so that you can see whether you might be stepping out into the path of a speeding car or heavy goods vehicle (HGV)?

To genuinely appreciate the value of threat prediction and modelling, it is important to understand the terms that contribute to the risk management practices.

2.1 What Is an Asset?

The term asset can have many different meanings so it is immensely helpful to look at some dictionary entries, origins and definitions so that the term can be effectively applied and understood against your business.

The Dictionary (Cambridge Dictionary, 2019a) describes an asset as being:

> Something having value, such as a possession or property, that is owned by a person, business, or organization.

Whereas the origins of the term asset go into further detail, by explaining that this relates to the sufficient estate and to anything associated to property (Online Etymology Dictionary, 2022)

> 1530s, "sufficient estate," from Anglo-French assetz, asetz (singular), from Old French assez "sufficiency, satisfaction; compensation" (11c.), noun use of adverb meaning "enough, sufficiently; very much, a great deal," from Vulgar Latin *ad satis "to sufficiency," from Latin ad "to" (see ad-) + satis "enough," from PIE root *sa- "to satisfy".
>
> At first a legal word meaning "sufficient estate" (to satisfy debts and legacies), it passed into a general sense of "property," especially "any property that theoretically can be converted to ready money" by 1580s. Figurative use from 1670s. Asset is a 19c. artificial singular. Corporate asset stripping is attested from 1972.

Further understanding can be gleaned from the numerous asset definitions that are available:

NIST SP 800-160, Vol. 2, Rev. 1. Developing Cyber-Resilient Systems (Ross et al., 2021)

> An asset refers to an item of value to stakeholders.
>
> Assets may be tangible (e.g., a physical item, such as hardware, firmware, computing platform, network device, or other technology component, or individuals in key or defined roles in organizations) or intangible (e.g., data, information, software, trademark, copyright, patent, intellectual property, image, or reputation).
>
> An item of value to stakeholders. An asset may be tangible (e.g., a physical item such as hardware, firmware, computing platform, network device, or other technology component) or intangible (e.g., humans, data, information, software, capability, function, service, trademark, copyright, patent, intellectual property, image, or reputation).
>
> The value of an asset is determined by stakeholders in consideration of loss concerns across the entire system life cycle. Such concerns include but are not limited to business or mission concerns.

NIST Interagency Report 7693. Specification for Asset Identification 1.1 (Wunder et al., 2011)

> Anything that has value to an organization, including, but not limited to, another organization, person, computing device, information technology (IT) system, IT network, IT circuit, software (both an installed instance and a physical instance), virtual computing platform (common in cloud and virtualized computing), and related hardware (e.g., locks, cabinets, keyboards).

NISTIR 8011, Volume 1. Automation Support for Security Control Assessments (Dempsey et al., 2017)

Resources of value that an organization possesses or employs.

NISTIR 8286. Integrating Cybersecurity and Enterprise Risk Management (ERM) (Stine et al., 2020)

The data, personnel, devices, systems, and facilities that enable the organization to achieve business purposes.

National Initiative for Cybersecurity Careers and Studies (NICCS) Cybersecurity Glossary (National Initiative for Cybersecurity Careers and Studies (NICCS))

A person, structure, facility, information, and records, information technology systems and resources, material, process, relationships, or reputation that has value.

Extended Definition: Anything useful that contributes to the success of something, such as an organizational mission; assets are things of value or properties to which value can be assigned.

Common to all these definitions is that an asset is something of value to an organization, whether this is tangible or intangible and that this should not limited to IT systems.

Consequently, when looking at your organization you should look at your business operations much the same as a medical person looks at the anatomy of the human body. They clearly know where all the critical and valuable organs exist and the importance of keeping their health and protecting them from harm/compromise.

2.2 *What Is a Threat?*

As per the earlier section to understand the term asset, we can glean some valuable insights by looking at the dictionary entries, origins and definitions that are available to us.

The Dictionary (Cambridge Dictionary, 2019b) entry describes a threat as being:

> The possibility that something unwanted will happen, or a person or thing that is likely to cause something unwanted to happen.

The origins of the term threat (Etymology Online Dictionary, 2022a) references an association to oppression, coercion or menace:

> Old English þreat "crowd, troop," also "oppression, coercion, menace," related to þreotan "to trouble, weary," from Proto-Germanic *thrautam (source also of Dutch verdrieten, German verdrießen "to vex"), from PIE *treud- "to push, press squeeze" (source also of Latin trudere "to press, thrust," Old Church Slavonic trudu "oppression," Middle Irish trott "quarrel, conflict,"' Middle Welsh cythrud "torture, torment, afflict"). Sense of "conditional declaration of hostile intention" was in Old English.

Table 1 Type of threats

1. Traditional	This type of threat originates from deliberate actions from an individual or group that have the intention of causing harm to the intended victim(s) and can be summarized using the **TESS(SS)** (The additional 'SS' has been added to represent global government's) **OC** (NATO Standardization Office (NSO), 2015)	**T**errorism **E**spionage **S**abotage **S**ubversion **S**tate **S**ponsored Cyber-Attacks **O**rganized **C**rime
2. Non-traditional	This type of threat originates from non-deliberate (accidental/non-malicious) actions of an individual or group that results in an impactful event against another.	Theft Natural Disasters (e.g. Fire, Flood, Earthquake, etc.) Employee error Amateur hacker

A choice of industry sources supplies the following definitions:

National Institute of Standards Technology Glossary (NIST, 2015)

Any circumstance or event with the potential to adversely impact organizational operations (including mission, functions, image, or reputation), organizational assets, individuals, other organizations, or the Nation through an information system via unauthorized access, destruction, disclosure, modification of information, and/or denial of service.

The IETF Trust Internet Security Glossary, Version 2 (Shirey, 2007)

A potential for violation of security, which exists when there is an entity, circumstance, capability, action, or event that could cause harm.

IEFT Internet Security Glossary RFC 2828 (Shirey, 2013) – ***Obsoleted by RFC 4949***

A potential for violation of security, which exists when there is a circumstance, capability, action, or event that could breach security and cause harm.

Threats to a business present themselves in several ways but typically these can be categorized into the following two types of threat, as shown in Table 1.

To better evaluate the potential likelihood of a threat materializing, you need to understand both the capabilities of any threat actors and your organization's weaknesses (aka Vulnerabilities) to such threats.

2.3 *What Is a Vulnerability?*

The dictionary (Cambridge Dictionary. VULNERABILITY | Meaning in the Cambridge English Dictionary) describes a vulnerability as being:

> The quality of being vulnerable (= able to be easily hurt, influenced, or attacked), or something that is vulnerable.

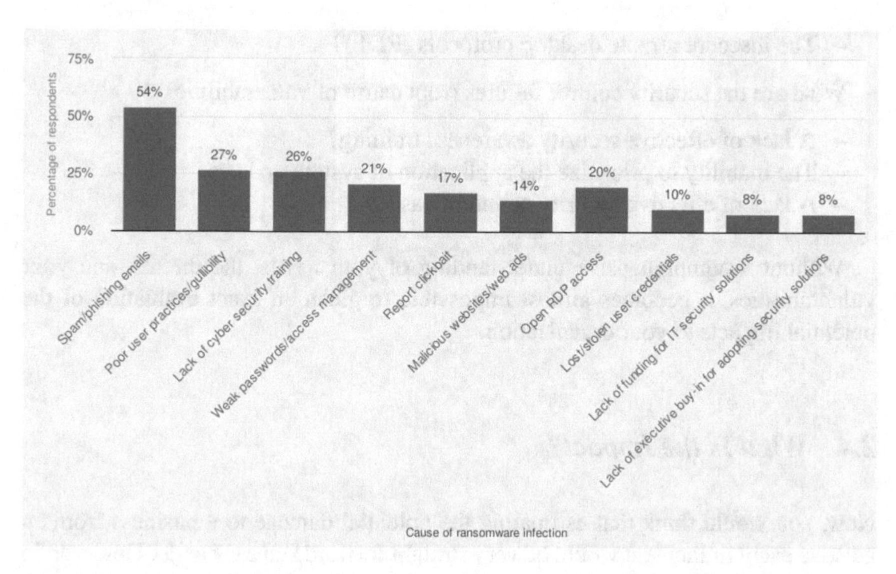

Fig. 1 Most common delivery methods and cybersecurity vulnerabilities causing ransomware infections according to MSPs worldwide as of 2020

The origin of this term is taken from the adjective 'Vulnerable' (Etymology Online Dictionary, 2022b):

> c. 1600, from Late Latin vulnerabilis "wounding," from Latin vulnerare "to wound, hurt, injure, maim," from vulnus (genitive vulneris) "wound," perhaps related to vellere "pluck, to tear" (see svelte), or from PIE *wele-nes-, from *wele- (2) "to strike, wound" (see Valhalla).

The national institute of standards and technology (NIST) define a vulnerability as being (CSRC, 2015):

> A weakness in an information system, system security procedures, internal controls, or implementation that could be exploited or triggered by a threat source.

Organization often misuse or confuse the terms threat and vulnerability (Hell, 2021) and can often refer to a security control failure as being a vulnerability when it might be the cause of a vulnerability.

Imagine trying to evaluate a ransomware risk scenario, without understanding what the threat and vulnerabilities are (as depicted in Fig. 1 (Statistica, 2021)).

- Is Ransomware the threat or is it the organized criminal gangs that are using the ransomware threat (e.g. LV, Conti, Snatch, Alphav (BlackCat), BlackByte, LockBit 2.0, etc. (Ransomware Database, 2022)?
- What are the vulnerabilities?
  - The unpatched systems?
  - The end user, who accidentally clicks on the malicious email link?
  - The insecure systems?

  - The insecure remote desktop protocols (RDP)?
- What are the security control failures (root cause of vulnerabilities)?
  - A lack of effective security awareness training?
  - The inability to prioritize the application of system updates.
  - A lack of effective security countermeasures?

Without a comprehensive understanding of your assets, the threats, and your vulnerabilities, it becomes almost impossible to glean an exact evaluation of the potential impacts to your organization.

2.4 *What Is the Impact?*

Now, you would think that estimating the potential damage to a business from an adverse event or incident would be very straight forward and easy to do. However, in truth, this is far from being the case, with many businesses applying a more laissez-faire approach to their impact analysis using reference tables, such as that shown in Table 2.

When using such a reference table, you are likely to see a varied interpretation of the impact analysis and with the reality being that carrying out effective impact analysis has far more variables to it and, like the weather, this can be very unpredictable.

For example, how do you accurately forecast the impact on customers if you do not know when an adverse event or impact may occur?

- Do you go for the worst-case scenario *(which may appear to be unrealistic)*, supply an optimistic best case *(again may appear unrealistic)* or do you go for the average?
- What happens when the impact analysis has varying scores across multiple impact areas?
  - Do you use the worst-case score?
  - Do you use the best-case score?
  - Do you use an average from all the impact areas?

Here, we may have only addressed the primary losses but what about any impactful secondary losses, for example how secondary stakeholders may adversely react to a primary loss event. Suddenly, impact analysis does not appear to be as straight forward as it first seemed and becomes increasingly difficult if your threat predictions and modelling practices do not help to paint a correct picture of the threats.

Table 2 Qualitative impact analysis reference table

Impact level		Impact areas				
Impact rating	General description	Effect on customers	Financial cost	Health and safety	Damage to to reputation	Legal, contractual & organizational compliance
1	Negligible	No effect	Very little or none	Very small additional risk	Negligible	No implications
2	Slight	Some local disturbance to normal business operations	Some	Within acceptable limits	Slight	Small risk of not meeting compliance
3	Moderate	Can still deliver product/service with some difficulty	Unwelcome but could be borne	Elevated risk requiring immediate attention	Moderate	In definite danger of operating illegally
4	High	Business is crippled in key areas	Severe effect on income and/or profit	Significant danger to life	High	Operating illegally in some areas
5	Very high	Out of business; no service to customers	Crippling; the organization will go out of business	Real or strong potential loss of life	Very high	Severe fines and possible imprisonment of staff

3 Threat Intelligence

It is extremely important for your business to take the various threat information feeds (both free and paid-for resources) to help you understand the threats that might affect your organization and its vital assets/processes. However, not all threat information will be of value to your organization and, as a result, you need to analyse (filter) this information to help turn this into intelligence.

When investigating the concept of cyber threat intelligence (CTI) Juhani Matilainen (2021) explains that:

> As a concept, it is hard to get exact and pervasive definition on what is cyber threat intelligence (CTI). This is because many academics and professional literature in the field defines this term differently. Situation becomes even more confusing, when many commercial organisations enlist different product as "threat intelligence" (TI) (Tounsi & Rais, 2018).

Whilst employed in the Royal Air Force Police, during numerous overseas deployments on counter intelligence field team (CIFT) operations, (Colley & Development, Concepts and Doctrine Centre, 2011, 2–16), my role was to develop a network of human intelligence resources from both the military, contractors and third-country and local country personnel. Using soft skills, the CIFT would seek to obtain relevant and pertinent information that could be analysed and recorded in the IBM I2 intelligence platform (Questys Solutions, 2022), as depicted in Fig. 2.

Additionally, my role was to complete concise 5X5 intelligence reports so that any relevant information could be communicated both up the chain of command and out to the other coalition forces, so that the deployed military forces were 'Forewarned is being Forearmed'.

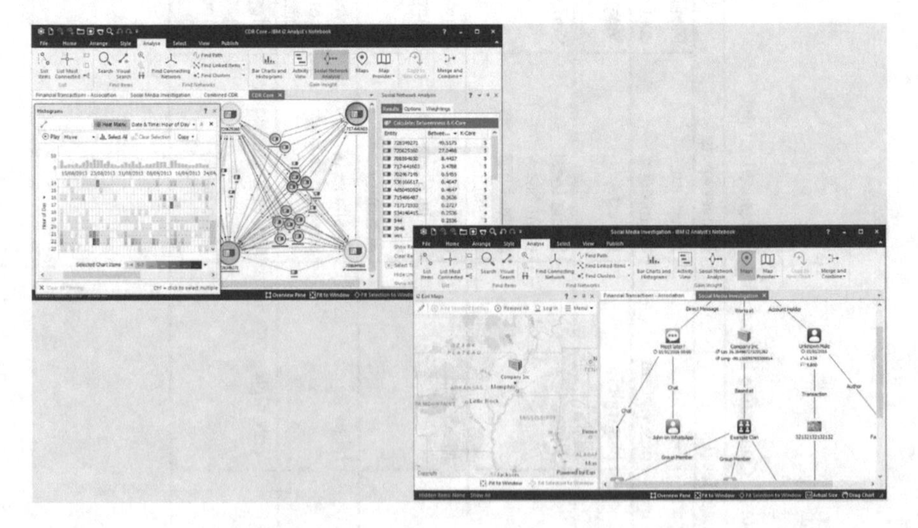

Fig. 2 I2 Intelligence platform

During a typical day, the CIFT could easily carry out an average of 5 interviews each day, resulting in around 1000 interviews being conducted over a 6-month deployment. However, only 400–500 of these interviews resulted in intelligence feeds into the I2 intelligence database or being recorded in the completion of a 5 × 5 intelligence report.

The same principle should apply to your threat intelligence practices. Avoid the urge to regurgitate the wealth of threat information that is available to you and look to find useful threat intelligence resources that are relevant and can be aligned to your business' valued assets/operations.

The results of your threat intelligence feeds should support the organization's risk management practices, and which help your leadership team to understand and appreciate what is important for the safeguarding of your business and which help to drive informed decision-making activities.

4 Developing Your Threat Prediction and Modelling Capabilities

Having found the information resources that you are going to use, next, you need to start to analyse the information so that the intelligence can be presented in a way that it clearly conveys the prevalence of any perceived threats in a concise manner. Think of it as looking at your business through the eyes of one of your threat actors (remembering that these can be both traditional and non-traditional).

Sun Tzu, the ancient Chinese military general, strategist and philosopher is quoted as saying (Tzu, 2019, 3:6:1–6):

> Know yourself and know your enemy.
> You will be safe in every battle.
> You may know yourself but not know the enemy.
> You will then lose one battle for every one you win.
> You may not know yourself or the enemy.
> You will then lose every battle.

Centuries ago, Sun Tzu found the importance of knowing your establishments' strengths and weaknesses (Asset & Vulnerability Management) whilst understanding your threats, and the same ethos still rings true for today's protection of Artificial Intelligence (AI) and National Security operations.

Consequently, you need to evaluate your business estate (especially prioritizing the high-value assets/operations) against your threat actors. This should be done for both the traditional threats and the non-traditional threats.

4.1 Traditional Threats

Within the traditional threats, you need to look at your business through the eyes of an aggressor.

- Terrorism (CPNI, 2014).
  - Could your business operations be impacted by a terrorist attack?
  - Are your valued business assets/operations sited near (or reliant upon) a high-risk terrorist target?
  - Do your employees rely on transport hubs that could be the target of terrorism?
  - If your employees must travel for work, do they travel in a group (e.g. on the same plane, train, etc.)?
  - Do you understand the tactics and techniques that may be used by a terrorist group?
  - What contingency plans do you have in place to help protect your business from being affected by terrorist activities?
- Espionage (CPNI, 2020).

  I was once told by a director of a global manufacturing company that their business was not concerned about any form of espionage. This was even though they were constantly innovating, inventing and bringing new products and services to market and have over 100 registered patents (Justia Patents, 2021), which had helped them to become a highly successful and profitable company.

  Imagine the value this industrial espionage (Beattie, 2021) would present to a rival organization, helping them to shorten the development time by using stolen data to speed up their development of new products, so that products could be patented and brought to market ahead of the company that came up with the product idea, in the first place.

  - Do you know the value of your intellectual property (IP), both to yourselves and one (or more) of your competitors?
  - Do you understand how a competitor might gain unauthorized access to your IP?
  - Do you appreciate how your IP might become compromised?
- Sabotage (Dictionary.com, 2022)

  We all hope that all our employees will 'Never Bite the Hand That Feeds Them!' and will remain 100% loyal to their employees and that your business would never become the target of protestors, who might use sabotage to convey their messages.

  However, that is rarely the truth, with such things being reported:

  - Over one third of all insider incidents are caused by malicious insiders (Wadhwani, 2022).
  - In 2020, 92 percent of oil spills were caused by sabotage and theft – like previous years (Ejoh, 2021).

  - The FBI in Washington State has been investigating at least 41 incidents of eco-sabotage (Spencer, 2021).

- **S**ubversion (Soetrust, 2021).

  It is likely that a business' employees are unlikely to be happy and content all the time, and they are likely to be instances of rebellion and disturbance, caused by their employees. However, it is important that these do not affect the valuable business operations. Consequently, it is important to enable communication channels between the business leadership and management teams, and the employees.
- **S**tate **S**ponsored Cyber-Attacks (Cybersecurity & Infrastructure Security Agency, 2022).

  I recently added State Sponsored (SS) cyber-attacks to the list of traditional (TESSOC) threats, as these threat actors are deliberately trying to target their victims and have an impact on their enemies. This type of threat has become so prevalent that it has been reported that 50% of the United States technology executives perceive this as being the greatest threat (de León, 2020).

  These groups tend to be well organized, highly skilled and well-funded and who seek to have an economical, financial or financial impact on the enemies of the states that are sponsoring these cyber-attacks.
- **O**rganized **C**rime (NCSC, 2017).

  The modern criminal gangs have recognized and embraced the profits that can be made from extracting company's sensitive data or extorting an organization, by making their IT systems or valued data assets unavailable (e.g. Ransomware).

4.2 Non-traditional Threats

This is one of the most difficult aspects to forecast and guess upon. How likely is it that your business is likely to become a victim of theft, suffer an earthquake, flood or fire, suffer an end-user making a mistake, an employee being negligent or becoming the victim of an amateur hacker (script kiddy)?

In areas where there is a higher crime rate or earthquakes, it might be possible to have an increased insight for predicting the potential likelihood of such an event. However, employee negligence or the erratic actions of a script kiddy is far more difficult to estimate.

5 Threat Modelling

Threat modelling is the structured process for finding and counting potential threats to help an organization prioritize their security mitigations. The aim of threat

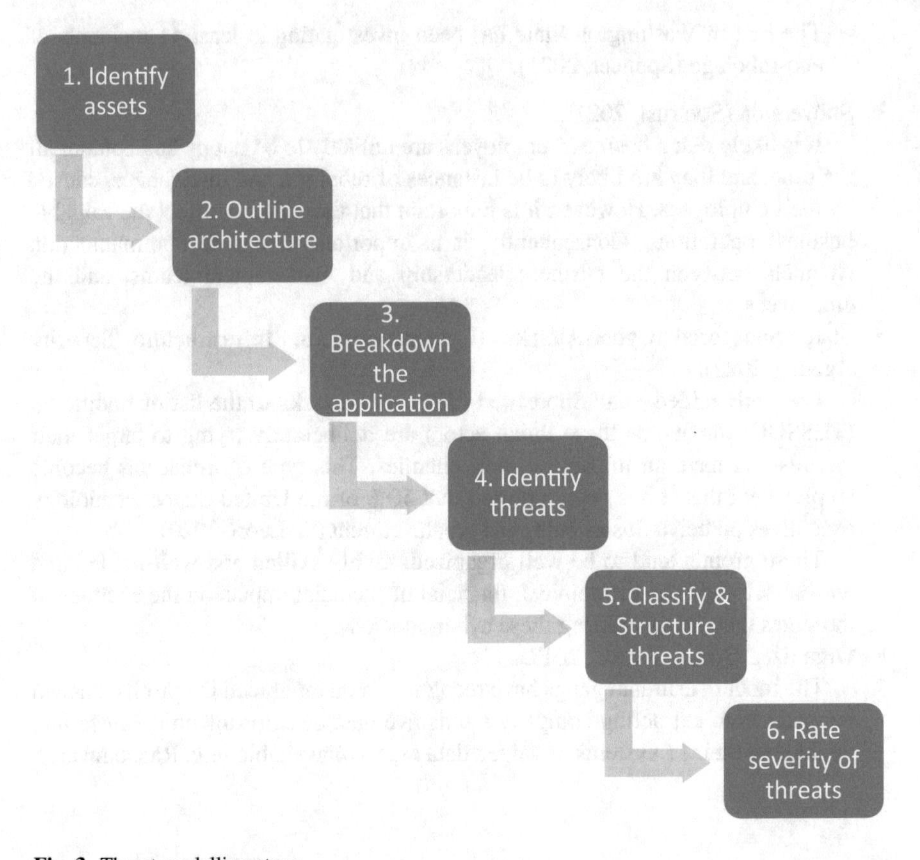

Fig. 3 Threat modelling steps

modelling is to help defenders and their security teams to be better equipped with an analysis of what the identified threat landscape are, the most likely attacks, their method, motive, and target system and what security controls are needed.

Typically, there are six main steps to threat modelling, as depicted in Fig. 3 (EC Council).

In addition, when creating/developing your threat modelling method it is extremely important that you understand the various available models, so that you can adopt the most suitable for your organization, e.g.

- **STRIDE** (jegeib, 2022)
  - **S**poofing
    Involves illegally accessing and then using another user's authentication information, such as username and password.
  - **T**ampering
    Involves the malicious modification of data. Examples include unauthorized changes made to persistent data, such as that held in a database, and the alteration of data as it flows between two computers over an open network, such as the Internet.

- **R**epudiation
  Associated with users who deny performing an action without other parties having any way to prove otherwise—for example, a user performs an illegal operation in a system that lacks the ability to trace the prohibited operations. Non-Repudiation refers to the ability of a system to counter repudiation threats. For example, a user who purchases an item might have to sign for the item upon receipt. The vendor can then use the signed receipt as evidence that the user did receive the package.
- **I**nformation disclosure
  Involves the exposure of information to individuals who are not supposed to have access to it—for example, the ability of users to read a file that they were not granted access to, or the ability of an intruder to read data in transit between two computers.
- **D**enial of Service (DoS)
  Denial of service (DoS) attacks deny service to valid users—for example, by making a Web server temporarily unavailable or unusable. You must protect against certain types of DoS threats simply to improve system availability and reliability.
- **E**levation of privilege
  An unprivileged user gains privileged access and thereby has sufficient access to compromise or destroy the entire system. Elevation of privilege threats include those situations in which an attacker has effectively penetrated all system defenses and become part of the trusted system itself, a dangerous situation indeed.

• The **P**rocess for **A**ttack **S**imulation **and** **T**hreat **A**nalysis **(PASTA)** (Welekwe, 2021)

  A seven-step, attack-centric methodology designed in 2015 to help organizations align technical requirements with business objectives while considering business impact analysis and compliance requirements. The goal of this methodology is to provide a dynamic threat identification, enumeration, and scoring process. PASTA is focused on guiding teams to identify, count, and prioritize threats dynamically.
  The overall sequence is as follows:

  1. Define business objectives
  2. Define technical scope
  3. Application decomposition
  4. Threat analysis
  5. Vulnerability and weaknesses analysis
  6. Attack enumeration and modeling
  7. Risk analysis and countermeasures.

• **Trike** (trikers, 2022).
  Trike is an open-source threat modeling methodology and tool. The project began in 2006 as an attempt to improve the efficiency and effectiveness of existing threat modeling methodologies and is being actively used and developed.

- Defining a system – Requirement Model
  - Risk Assessment – CRUD
    - Creating
    - Reading
    - Updating
    - Deleting
  - Data Flow Diagram (DFD)
  - Assigning Risk Values

- **Visual, Agile and Simple Threat (VAST)**, (GeeksforGeeks, 2021)

  The methodology provides actionable outputs for the unique needs of various stakeholders like application architects and developers, cyber security personnel, etc. It provides a unique application and infrastructure visualization scheme such that the creation and use of threat models do not require specific security subject matter expertise.

  - Automation
    - Eliminate repetition
    - Ongoing
    - Scaled to encompass entire Enterprise
  - Integration
    - Integration with tools
    - Support Agile DevOps
  - Collaboration
    - Key stakeholders:
      - Software developers
      - Systems architects
      - Security team
      - Senior executives

- **Damage Reproducibility, Exploitability, Affected users and Discoverability (DREAD)** (Weston, 2021)

  This is a classification scheme for determining and comparing the amount of risk related to each identified threat. In using the DREAD model, a threat modeling team can quantify, or calculate, a numeric value for the security risk provided by each threat.
  - **D**amage potential
    The damage caused when the threat exploits?
  - **R**eproducibility
    How easily the threat can be exploited?
  - **E**xploitability
    The frequency or size of the vector related to the threat?
  - **A**ffected users
    How wide the impact of the threat would be felt if realized?
  - **D**iscoverability
    How easy it is to discover the threat?

- **Operationally Critical Threat, Asset and Vulnerability Evaluation (OCTAVE)** (Luo et al., 2021)

  CERT/CC (Computer Emergency Response Team/Coordination Center) released OCTAVE in 1999. The OCTAVE method has become one of the mainstream TARA methods in the world.

  The OCTAVE methodology is an approach that divides the assessment into three phases in which management issues and technical issues are examined and discussed so that the organization's staff can take full ownership of the organization's information security needs.

The OCTAVE method is characterized as an assessment approach that combines assets, threats, and vulnerabilities. It allows managers to use the results of the assessment to determine the OCTAVE method, which is characterized by a combination of asset, threat, and vulnerability assessments.

In addition, managers can use the results of the assessment to prioritize risks to be addressed. It also incorporates how the computing infrastructure is used and its role in achieving the organization's business objectives. OCTAVE is integrated with the interrelated technical aspects of computing infrastructure configuration.

It also allows for a flexible, customizable, and repeatable approach that can be customized according to the needs of different organizations.

- **Threat Attack Trees** (Saini et al., 2008), as depicted at Fig. 4 (Clark, 2020) Attacks trees were defined by Bruce Schneier to model threats against computer systems. By understanding all the different ways in which a system can be attacked, we can likely design countermeasures to thwart those attacks. Further, by understanding who the attackers are – not to mention their abilities, motivations, and goals – maybe we can install the proper countermeasures to deal with the real threats.

  Attack Trees provide a formal, methodical way of describing the security of systems, based on varying attacks. A tree structure is used represent attacks against a system, with the goal as the root node and different ways of achieving that goal as leaf nodes. Each node becomes a subgoal, and children of that node are ways to achieve that subgoal. OR nodes are used to represent alternatives and AND nodes are used to represent different steps toward achieving the same goal.

  Once the tree is built, one can assign values to the various leaf nodes, then make calculations about the nodes. Once the values are assigned, one can calculate the security of the goal. The attack attributes assist in associating risk with an attack.

  An Attack Tree can include special knowledge or equipment that is needed, the time required to complete a step, and the physical and legal risks assumed by the attacker. The values in the Attack Tree could also be operational or development expenses. An Attack Tree supports design and requirement decisions. If an attack costs the perpetrator more than the benefit, that attack will most likely not occur. However, if there are easy attacks that may result in benefit, then those need a defense.

You will find that you may need to adopt one, or more, of these threat models to articulate the threats as an intelligence feed into the security risk management

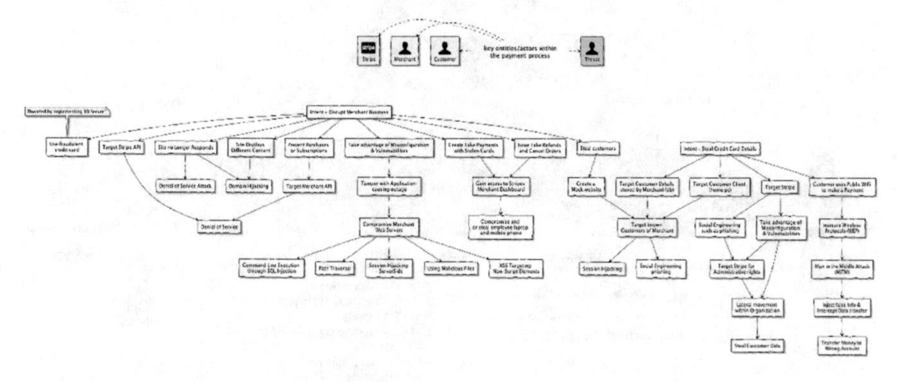

Fig. 4 Online payment threat attack tree

practice. An essential aspect of the threat modelling is the creation of correct and relevant threat scenarios.

5.1 Threat Scenarios

An essential part of threat prediction and modelling is the creation of realistic and plausible threat scenarios. The national institute of standards and technology (NIST) (Editor, CSRC Content) define threat scenarios as being:

> A set of discrete threat events, associated with a specific threat source or multiple threat sources, partially ordered in time.

Your threat scenarios help you to contextualize any identified threats against your valued business operations to aid you to show how such threats are relevant to your organization.

Having created a library of threat scenarios, you are then better equipped to carry out analysis of possible future events by considering several alternative possible outcomes. This analysis then enables better informed decision making through the consideration of the possible outcomes and their potential implications.

The output from your threat scenarios and analysis exercises provides essential input to enhance your security risk management practices, with the risks being assessed from plausible and relevant threat scenarios.

An example of a pictorial version of a cyber-threat scenario is provided at Fig. 5 and a tabular version of a cyber-threat scenario is provided at Table 3 (Kim & Cha, 2011).

Threat scenarios should not be limited to just the 'Cyber' domain and should address both the traditional and non-traditional aspects and should be created from

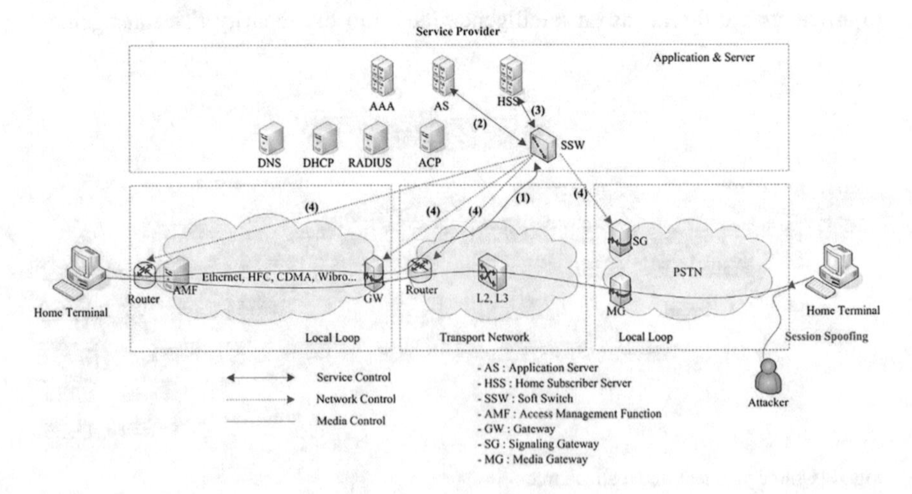

Fig. 5 Broadband convergence network

Table 3 Modification of session message

Use-case name	Modification of session message				
Threat actor(s)	Home terminal attacker				
Security risk property(ies)	Confidentiality, integrity				
Description	Home terminal attacker breaks confidentiality and integrity of user service by modification of session message using session spoofing				
Target	Network protocol between home terminal and local loop, private user information, such as banking and currency.				
Expected damage	Fabrication of user's accounting, financial data and private information				
Pre-conditions	–				
Post-conditions	–				
	No.	**Scenario**	**Impact**	**Sensitivity**	**Relationship**
Threat scenario	1.	The home terminal attacker performs a spoofing attack against the connected session between the home terminal and the local loop, using network scanning tools.	Medium	High	–
	2.	The home terminal attacker communicates with an application server, using the faked message against the spoofed session of the home user.	High	Medium	–
Consideration	Fabrication of the connected session requires advanced techniques, such as understanding the application service layer, and attack skills and a fake service system.				
	However, if this kind of attack is not protected by encryption of the session message in the application layer, there is no appropriate safeguard.				
Countermeasure	Packets with an unauthorized IP address are blocked in the SSW.				
	Only an authorized user is allowed to service control equipment through user authentication of the request service.				
	In case of protecting traffic analysis in a specific BcN service, the service supports encryption of control messages in the home terminal.				

both actual real-life events/incidents and from industry threat intelligence feeds, as depicted in Table 4 (Blank & Gallagher, 2012).

6 Using the Mitre ATT&CK (MITRE. MITRE ATT&CK®) and CAPEC (MITRE. CAPEC – Common Attack Pattern Enumeration and Classification (CAPECTM)) for Cyber Threat Prediction and Modelling

Mitre have supplied a both the ATT&CK and Common Attack Pattern Enumerations and Classifications (CAPEC™) frameworks to help organizations to better appreciate the tactics and techniques that have been identified as being used by known Advanced Persistent Threat (APT), (ISACA, 2013) groups.

By having better insights into the tactics and techniques that might be used against you, your organization will be able to make better, improved, risk decisions about their most suitable and proper risk responses.

For example, the MITRE ATT&CK framework has named the following 14 tactics for an Enterprise environment (MITRE, 2022a) and 92 tactics for Mobile environments (MITRE, 2022b), as depicted in Table 5:

In addition to presenting a comprehensive list of known tactics and techniques (Enterprise & Mobile), it also provides you with some suggested mitigation measures (MITRE, 2022c, d) e.g.

- **T1087 – Account Discovery** (MITRE, 2022e).
  - **M1028 – Operating System Configuration** (MITRE, 2022f).
    - Prevent administrator accounts from being enumerated when an application is elevating through UAC since it can lead to the disclosure of account names.
      The Registry key is located HKLM\ SOFTWARE\Microsoft\Windows\ CurrentVersion\Policies\CredUI\EnumerateAdministrators.
      It can be disabled through GPO: Computer Configuration > [Policies] > Administrative Templates > Windows Components > Credential User Interface: E numerate administrator accounts on elevation.

As a financial services organization, you can drill down into these tactics to understand how these tactics might be leveraged against your organization and by which APT groups (MITRE. Groups | MITRE ATT&CK®), e.g. **G0018 – admin@338** (MITRE, 2022g).

The Cyber Kill Chain is depicted at Fig. 6 (MITRE, 2022h).

The Cyber Kill Chain can then be further filtered to represent your business environment, e.g. an environment that uses MS Windows, Office 365 and Azure AD, as depicted in Fig. 7.

Table 4 Inputs: Threat source identification

	Provided to		
	Tier 1	Tier 2	Tier 3
From Tier 1: (Organization level) Sources of threat information deemed to be credible (e.g. open source and/or classified threat Reports, previous risk/threat assessments). (Section 3.1, Task 1–4) Threat source information and guidance specific to Tier 1 (e.g. threats related to organizational governance, core missions/business functions, management/operational policies, procedures, and structures, external mission/business relationships). Taxonomy of threat sources, annotated by the organization, if necessary. (Table D-2) Characterization of adversarial and non-adversarial threat sources. Assessment scales for assessing adversary capability, intent, and targeting, annotated by the organization, if necessary. (Table D-3, Table D-4, Table D-5) Assessment scale for assessing the range of effects, annotated by the organization, if necessary. Threat sources identified in previous risk assessments, if appropriate.	No	Yes	Yes
From Tier 2: (Mission/business process level) Threat source information and guidance specific to Tier 2 (e.g. threats related to mission/business processes, EA segments, common infrastructure, support services, common controls, and external dependencies). Mission/business process-specific characterization of adversarial and non-adversarial threat sources.	Yes via RAR	Yes, via peer sharing	Yes
From Tier 3: (Information system level) Threat source information and guidance specific to Tier 3 (e.g. threats related to information systems, information technologies, information system components, applications, networks, environments of operation). Information system-specific characterization of adversarial and non-adversarial threat sources.	Yes via RAR	Yes via RAR	Yes, via peer sharing

Table 5 MITRE ATT&CK tactics list

Enterprise			Mobile		
ID	Name	Description	ID	Name	Description
TA0043	Reconnaissance	The adversary is trying to gather information they can use to plan future operations.	T1435	Access calendar entries	An adversary could call standard operating system APIs from a malicious application to gather calendar entry data, or with escalated privileges could directly access files containing calendar data.
TA0042	Resource development	The adversary is trying to establish resources they can use to support operations.	T1433	Access call Log	On Android, an adversary could call standard operating system APIs from a malicious application to gather call log data, or with escalated privileges could directly access files containing call log data.
TA0001	Initial access	The adversary is trying to get into your network.	T1432	Access contact List	An adversary could call standard operating system APIs from a malicious application to gather contact list (i.e., address book) data, or with escalated privileges could directly access files containing contact list data.
TA0002	Execution	The adversary is trying to run malicious code.	T1517	Access notifications	A malicious application can read notifications sent by the operating system or other applications, which may contain sensitive data such as one-time authentication codes sent over SMS, email, or other mediums. A malicious application can also dismiss notifications to prevent the user from noticing that the notifications arrived and can trigger action buttons contained within notifications.
TA0003	Persistence	The adversary is trying to maintain their foothold.	T1413	Access sensitive data in device logs	On versions of Android prior to 4.1, an adversary may use a malicious application that holds the READ_LOGS permission to obtain private keys, passwords, other credentials, or other sensitive data stored in the device's system log. On Android 4.1 and later, an adversary would need to attempt to perform an operating system privilege escalation attack to be able to access the log.

TA0005	Defence evasion	The adversary is trying to avoid being detected.	T1409	Access Stored Application Data	Adversaries may access and collect application data resident on the device. Adversaries often target popular applications such as Facebook, WeChat, and Gmail.
TA0006	Credential access	The adversary is trying to steal account names and passwords.	T1438	Alternate network mediums	Adversaries can communicate using cellular networks rather than enterprise Wi-Fi in order to bypass enterprise network monitoring systems. Adversaries may also communicate using other non-Internet Protocol mediums such as SMS, NFC, or Bluetooth to bypass network monitoring systems.
TA0007	Discovery	The adversary is trying to figure out your environment.	T1418	Application discovery	Adversaries may seek to identify all applications installed on the device. One use case for doing so is to identify the presence of endpoint security applications that may increase the adversary's risk of detection. Another use case is to identify the presence of applications that the adversary may wish to target.
TA0008	Lateral movement	The adversary is trying to move through your environment.	T1427	Attack PC via USB connection	With escalated privileges, an adversary could programme the mobile device to impersonate USB devices such as input devices (keyboard and mouse), storage devices, and/or networking devices in order to attack a physically connected PC this technique has been demonstrated on Android. We are unaware of any demonstrations on iOS.
TA0009	Collection	The adversary is trying to gather data of interest to their goal.	T1402	Broadcast receivers	An intent is a message passed between Android application or system components. Applications can register to receive broadcast intents at runtime, which are system-wide intents delivered to each app when certain events happen on the device, such as network changes or the user unlocking the screen. Malicious applications can then trigger certain actions within the app based on which broadcast intent was received.
TA0011	Command and control	The adversary is trying to communicate with compromised systems to control them.	T1616	Call control	Adversaries may make, forward, or block phone calls without user authorization. This could be used for adversary goals such as audio surveillance, blocking or forwarding calls from the device owner, or C2 communication.

(continued)

Table 5 (continued)

Enterprise			Mobile		
ID	Name	Description	ID	Name	Description
TA0010	Exfiltration	The adversary is trying to steal data.	T1429	Capture audio	Adversaries may capture audio to collect information on a user of a mobile device using standard operating system APIs. Adversaries may target audio information such as user conversations, surroundings, phone calls, or other sensitive information.
TA0040	Impact	The adversary is trying to manipulate, interrupt, or destroy your systems and data.	T1512	Capture camera	Adversaries may utilize the camera to capture information about the user, their surroundings, or other physical identifiers. Adversaries may use the physical camera devices on a mobile device to capture images or video. By default, in Android and iOS, an application must request permission to access a camera device which is granted by the user through a request prompt. In Android, applications must hold the android.permission.CAMERA permission to access the camera. In iOS, applications must include the NSCameraUsageDescription key in the Info.plist file and must request access to the camera at runtime.
			T1414	Capture clipboard data	Adversaries may abuse clipboard manager APIs to obtain sensitive information copied to the global clipboard. For example, passwords being copy-and-pasted from a password manager app could be captured by another application installed on the device.
			T1412	Capture SMS sessages	A malicious application could capture sensitive data sent via SMS, including authentication credentials. SMS is frequently used to transmit codes used for multi-factor authentication.
			T1448	Carrier billing fraud	A malicious app may trigger fraudulent charges on a victim's carrier billing statement in several different ways, including SMS toll fraud and SMS shortcodes that make purchases.

	T1510	Clipboard modification	Adversaries may abuse clipboard functionality to intercept and replace information in the Android device clipboard. Malicious applications may monitor the clipboard activity through the Clipboard Manager. On primary clip changed listener interface on Android to determine when the clipboard contents have changed. Listening to clipboard activity, reading the clipboard contents, and modifying the clipboard contents requires no explicit application permissions and can be performed by applications running in the background, however, this behaviour has changed with the release of Android 10.
	T1540	Code injection	Adversaries may use code injection attacks to implant arbitrary code into the address space of a running application. Code is then executed or interpreted by that application. Adversaries utilizing this technique may exploit capabilities to load code in at runtime through dynamic libraries.
	T1605	Command-line interface	Adversaries may use built-in command-line interfaces to interact with the device and execute commands. Android provides a bash shell that can be interacted with over the Android Debug Bridge (ADB) or programmatically using Java's Runtime package. On iOS, adversaries can interact with the underlying runtime shell if the device has been jail broken.
	T1436	Commonly used port	Adversaries may communicate over a commonly used port to bypass firewalls or network detection systems and to blend with normal network activity to avoid more detailed inspection.
	T1577	Compromise application executable	Adversaries may modify applications installed on a device to establish persistent access to a victim. These malicious modifications can be used to make legitimate applications carry out adversary tasks when these applications are in use.

(continued)

Table 5 (continued)

Enterprise			Mobile		
ID	Name	Description	ID	Name	Description
			T1532	Data encrypted	Data is encrypted before being exfiltrated in order to hide the information that is being exfiltrated from detection or to make the exfiltration less conspicuous upon inspection by a defender. The encryption is performed by a utility, programming library, or custom algorithm on the data itself and is considered separate from any encryption performed by the command and control or file transfer protocol. Common file formats that can encrypt files are RAR and zip.
			T1471	Data encrypted for impact	An adversary may encrypt files stored on the mobile device to prevent the user from accessing them, for example with the intent of only unlocking access to the files after a ransom is paid. Without escalated privileges, the adversary is generally limited to only encrypting files in external/shared storage locations. This technique has been demonstrated on Android. We are unaware of any demonstrated use on iOS.
			T1533	Data from local system	Sensitive data can be collected from local system sources, such as the file system or databases of information residing on the system.
			T1447	Delete device data	Adversaries may wipe a device or delete individual files in order to manipulate external outcomes or hide activity. An application must have administrator access to fully wipe the device, while individual files may not require special permissions to delete depending on their storage location.
			T1475	Deliver malicious App via authorized App Store	Malicious applications are a common attack vector used by adversaries to gain a presence on mobile devices. Mobile devices often are configured to allow application installation only from an authorized app store (e.g. Google Play Store or Apple App Store). An adversary may seek to place a malicious application in an authorized app store, enabling the application to be installed onto targeted devices.

	T1476	Deliver malicious App via other means	Malicious applications are a common attack vector used by adversaries to gain a presence on mobile devices. This technique describes installing a malicious application on targeted mobile devices without involving an authorized app store (e.g. Google Play Store or Apple App Store). Adversaries may wish to avoid placing malicious applications in an authorized app store due to increased potential risk of detection or other reasons. However, mobile devices often are configured to allow application installation only from an authorized app store which would prevent this technique from working.
	T1401	Device administrator permissions	Adversaries may request device administrator permissions to perform malicious actions.
	T1446	Device lockout	An adversary may seek to lock the legitimate user out of the device, for example to inhibit user interaction or to obtain a ransom payment.
	T1408	Disguise root/jailbreak indicators	An adversary could use knowledge of the techniques used by security software to evade detection. For example, some mobile security products perform compromised device detection by searching for particular artifacts such as an installed 'su' binary, but that check could be evaded by naming the binary something else. Similarly, polymorphic code techniques could be used to evade signature-based detection.
	T1520	Domain generation algorithms	Adversaries may use Domain Generation Algorithms (DGAs) to procedurally generate domain names for command-and-control communication, and other uses such as malicious application distribution.
	T1466	Downgrade to insecure protocols	An adversary could cause the mobile device to use less secure protocols, for example by jamming frequencies used by newer protocols such as LTE and only allowing older protocols such as GSM to communicate. Use of less secure protocols may make communication easier to eavesdrop upon or manipulate.

(continued)

Table 5 (continued)

Enterprise			Mobile		
ID	Name	Description	ID	Name	Description
			T1407	Download new code at runtime	An app could download and execute dynamic code (not included in the original application package) after installation to evade static analysis techniques (and potentially dynamic analysis techniques) used for application vetting or application store review.
			T1456	Drive-by compromise	As described by Drive-by Compromise, a drive-by compromise is when an adversary gains access to a system through a user visiting a website over the normal course of browsing. With this technique, the user's web browser is targeted for exploitation. For example, a website may contain malicious media content intended to exploit vulnerabilities in media parsers as demonstrated by the Android Stagefright vulnerability.
			T1439	Eavesdrop on insecure network communication	If network traffic between the mobile device and remote servers is unencrypted or is encrypted in an insecure manner, then an adversary positioned on the network can eavesdrop on communication.
			T1523	Evade analysis environment	Malicious applications may attempt to detect their operating environment prior to fully executing their payloads. These checks are often used to ensure the application is not running within an analysis environment such as a sandbox used for application vetting, security research, or reverse engineering. Adversaries may use many different checks such as physical sensors, location, and system properties to fingerprint emulators and sandbox environments. Adversaries may access android.os.SystemProperties via Java reflection to obtain specific system information. Standard values such as phone number, IMEI, IMSI, device IDs, and device drivers may be checked against default signatures of common sandboxes.

	T1428	Exploit enterprise resources	Adversaries may attempt to exploit enterprise servers, workstations, or other resources over the network. This technique may take advantage of the mobile device's access to an internal enterprise network either through local connectivity or through a Virtual Private Network (VPN).
	T1404	Exploit OS vulnerability	A malicious app can exploit unpatched vulnerabilities in the operating system to obtain escalated privileges.
	T1449	Exploit SS7 to redirect phone calls/SMS	An adversary could exploit signaling system vulnerabilities to redirect calls or text messages (SMS) to a phone number under the attacker's control. The adversary could then act as an adversary-in-the-middle to intercept or manipulate the communication. Interception of SMS messages could enable adversaries to obtain authentication codes used for multi-factor authentication.
	T1450	Exploit SS7 to track device location	An adversary could exploit signaling system vulnerabilities to track the location of mobile devices.
	T1405	Exploit TEE vulnerability	A malicious app or other attack vector could be used to exploit vulnerabilities in code running within the Trusted Execution Environment (TEE). The adversary could then obtain privileges held by the TEE potentially including the ability to access cryptographic keys or other sensitive data. Escalated operating system privileges may be first required in order to have the ability to attack the TEE. If not, privileges within the TEE can potentially be used to exploit the operating system.
	T1458	Exploit via charging station or PC	If the mobile device is connected (typically via USB) to a charging station or a PC, for example to charge the device's battery, then a compromised or malicious charging station or PC could attempt to exploit the mobile device via the connection.
	T1477	Exploit via radio interfaces	The mobile device may be targeted for exploitation through its interface to cellular networks or other radio interfaces.

(continued)

Table 5 (continued)

Enterprise			Mobile		
ID	Name	Description	ID	Name	Description
			T1420	File and directory discovery	On Android, command line tools or the Java file APIs can be used to enumerate file system contents. However, Linux file permissions and SELinux policies generally strongly restrict what can be accessed by apps (without taking advantage of a privilege escalation exploit). The contents of the external storage directory are generally visible, which could present concern if sensitive data is inappropriately stored there.
			T1541	Foreground persistence	Adversaries may abuse Android's start Foreground() API method to maintain continuous sensor access. Beginning in Android 9, idle applications running in the background no longer have access to device sensors, such as the camera, microphone, and gyroscope. Applications can retain sensor access by running in the foreground, using Android's startForeground() API method. This informs the system that the user is actively interacting with the application, and it should not be killed. The only requirement to start a foreground service is showing a persistent notification to the user.
			T1472	Generate fraudulent advertising revenue	An adversary could seek to generate fraudulent advertising revenue from mobile devices, for example by triggering automatic clicks of advertising links without user involvement.
			T1581	Geofencing	Adversaries may use a device's geographical location to limit certain malicious behaviours. For example, malware operators may limit the distribution of a second stage payload to certain geographic regions.

	T1617	Hooking	Adversaries may utilize hooking to hide the presence of artifacts associated with their behaviours to evade detection. Hooking can be used to modify return values or data structures of system APIs and function calls. This process typically involves using 3rd party root frameworks, such as Xposed or Magisk, with either a system exploit or pre-existing root access. By including custom modules for root frameworks, adversaries can hook system APIs and alter the return value and/or system data structures to alter functionality/visibility of various aspects of the system.
	T1417	Input capture	Adversaries may capture user input to obtain credentials or other information from the user through various methods.
	T1516	Input injection	A malicious application can inject input to the user interface to mimic user interaction through the abuse of Android's accessibility APIs.
	T1411	Input prompt	The operating system and installed applications often have legitimate needs to prompt the user for sensitive information such as account credentials, bank account information, or Personally Identifiable Information (PII). Adversaries may mimic this functionality to prompt users for sensitive information.
	T1478	Install insecure or malicious configuration	An adversary could attempt to install insecure or malicious configuration settings on the mobile device, through means such as phishing emails or text messages either directly containing the configuration settings as an attachment or containing a web link to the configuration settings. The device user may be tricked into installing the configuration settings through social engineering techniques.
	T1464	Jamming or denial of service	An attacker could jam radio signals (e.g. Wi-Fi, cellular, GPS) to prevent the mobile device from communicating.

(continued)

Table 5 (continued)

Enterprise			Mobile		
ID	Name	Description	ID	Name	Description
			T1579	Keychain	Adversaries may collect the keychain storage data from an iOS device to acquire credentials. Keychains are the built-in way for iOS to keep track of users' passwords and credentials for many services and features such as Wi-Fi passwords, websites, secure notes, certificates, private keys, and VPN credentials.
			T1430	Location tracking	An adversary could use a malicious or exploited application to surreptitiously track the device's physical location through use of standard operating system APIs.
			T1461	Lockscreen bypass	An adversary with physical access to a mobile device may seek to bypass the device's lockscreen.
			T1452	Manipulate App store rankings or ratings	An adversary could use access to a compromised device's credentials to attempt to manipulate app store rankings or ratings by triggering application downloads or posting fake reviews of applications. This technique likely requires privileged access (a rooted or jailbroken device).
			T1463	Manipulate device communication	If network traffic between the mobile device and a remote server is not securely protected, then an attacker positioned on the network may be able to manipulate network communication without being detected. For example, FireEye researchers found in 2014 that 68% of the top 1000 free applications in the Google Play Store had at least one Transport Layer Security (TLS) implementation vulnerability potentially opening the applications' network traffic to adversary-in-the-middle attacks.
			T1444	Masquerade as legitimate application	An adversary could distribute developed malware by masquerading the malware as a legitimate application. This can be done in two different ways: by embedding the malware in a legitimate application, or by pretending to be a legitimate application.

	T1403	Modify cached executable code	ART (the Android Runtime) compiles optimized code on the device itself to improve performance. An adversary may be able to use escalated privileges to modify the cached code in order to hide malicious behaviour. Since the code is compiled on the device, it may not receive the same level of integrity checks that are provided to code running in the system partition.
	T1398	Modify OS Kernel or boot partition	If an adversary can escalate privileges, he or she may be able to use those privileges to place malicious code in the device kernel or other boot partition components, where the code may evade detection, may persist after device resets, and may not be removable by the device user. In some cases (e.g. the Samsung Knox warranty bit as described under Detection), the attack may be detected but could result in the device being placed in a state that no longer allows certain functionality.
	T1400	Modify system partition	If an adversary can escalate privileges, he or she may be able to use those privileges to place malicious code in the device system partition, where it may persist after device resets and may not be easily removed by the device user.
	T1399	Modify trusted execution environment	If an adversary can escalate privileges, he or she may be able to use those privileges to place malicious code in the device's Trusted Execution Environment (TEE) or other similar isolated execution environment where the code can evade detection, may persist after device resets, and may not be removable by the device user. Running code within the TEE may provide an adversary with the ability to monitor or tamper with overall device behaviour.
	T1575	Native code	Adversaries may use Android's Native Development Kit (NDK) to write native functions that can achieve execution of binaries or functions. Like system calls on a traditional desktop operating system, native code achieves execution on a lower level than normal Android SDK calls.

(continued)

Table 5 (continued)

Enterprise			Mobile		
ID	Name	Description	ID	Name	Description
			T1507	Network information discovery	Adversaries may use device sensors to collect information about nearby networks, such as Wi-Fi and Bluetooth.
			T1423	Network service scanning	Adversaries may attempt to get a listing of services running on remote hosts, including those that may be vulnerable to remote software exploitation. Methods to acquire this information include port scans and vulnerability scans from the mobile device. This technique may take advantage of the mobile device's access to an internal enterprise network either through local connectivity or through a Virtual Private Network (VPN).
			T1410	Network traffic capture or redirection	An adversary may capture network traffic to and from the device to obtain credentials or other sensitive data, or redirect network traffic to flow through an adversary-controlled gateway to do the same.
			T1406	Obfuscated files or information	An app could contain malicious code in obfuscated or encrypted form, then deobfuscate or decrypt the code at runtime to evade many app vetting techniques.
			T1470	Obtain device cloud backups	An adversary who is able to obtain unauthorized access to or misuse authorized access to cloud backup services (e.g. Google's Android backup service or Apple's iCloud) could use that access to obtain sensitive data stored in device backups. For example, the Elcomsoft Phone Breaker product advertises the ability to retrieve iOS backup data from Apple's iCloud. Elcomsoft also describes obtaining WhatsApp communication histories from backups stored in iCloud.

	T1424	Process discovery	On Android versions prior to 5, applications can observe information about other processes that are running through methods in the Activity Manager class. On Android versions prior to 7, applications can obtain this information by executing the ps command, or by examining the /proc. directory. Starting in Android version 7, use of the Linux kernel's hidepid feature prevents applications (without escalated privileges) from accessing this information.
	T1604	Proxy through victim	Adversaries may use a compromised device as a proxy server to the Internet. By utilizing a proxy, adversaries hide the true IP address of their C2 server and associated infrastructure from the destination of the network traffic. This masquerades an adversary's traffic as legitimate traffic originating from the compromised device, which can evade IP-based restrictions and alerts on certain services, such as bank accounts and social media websites.
	T1544	Remote file copy	Files may be copied from one system to another to stage adversary tools or other files over the course of an operation. Files may be copied from an external adversary-controlled system through the Command-and-Control channel to bring tools into the victim network or onto the victim's device.
	T1468	Remotely track device without authorization	An adversary who is able to obtain unauthorized access to or misuse authorized access to cloud services (e.g. Google's Android Device Manager or Apple iCloud's Find my iPhone) or to an enterprise mobility management (EMM) / mobile device management (MDM) server console could use that access to track mobile devices.
	T1469	Remotely wipe data without authorization	An adversary who is able to obtain unauthorized access to or misuse authorized access to cloud services (e.g. Google's Android Device Manager or Apple iCloud's Find my iPhone) or to an EMM console could use that access to wipe enrolled devices.

(continued)

Table 5 (continued)

Enterprise			Mobile		
ID	Name	Description	ID	Name	Description
			T1467	Rogue cellular base station	An adversary could set up a rogue cellular base station and then use it to eavesdrop on or manipulate cellular device communication. A compromised cellular femtocell could be used to carry out this technique.
			T1465	Rogue Wi-Fi access points	An adversary could set up unauthorized Wi-Fi access points or compromise existing access points and, if the device connects to them, carry out network-based attacks such as eavesdropping on or modifying network communication.
			T1603	Scheduled task/job	Adversaries may abuse task scheduling functionality to facilitate initial or recurring execution of malicious code. On Android and iOS, APIs and libraries exist to facilitate scheduling tasks to execute at a specified date, time, or interval.
			T1513	Screen capture	Adversaries may use screen captures to collect information about applications running in the foreground, capture user data, credentials, or other sensitive information. Applications running in the background can capture screenshots or videos of another application running in the foreground by using the Android Media Projection Manager (generally requires the device user to grant consent). Background applications can also use Android accessibility services to capture screen contents being displayed by a foreground application. An adversary with root access or Android Debug Bridge (adb) access could call the Android screencap or screen record commands.
			T1451	SIM card swap	An adversary could convince the mobile network operator (e.g. through social networking, forged identification, or insider attacks performed by trusted employees) to issue a new SIM card and associate it with an existing phone number and account. The adversary could then obtain SMS messages or hijack phone calls intended for someone else.

	T1582	SMS control	Adversaries may delete, alter, or send SMS messages without user authorization. This could be used to hide C2 SMS messages, spread malware, or various external effects.
	T1437	Standard application layer protocol	Adversaries may communicate using a common, standardized application layer protocol such as HTTP, HTTPS, SMTP, or DNS to avoid detection by blending in with existing traffic.
	T1521	Standard cryptographic protocol	Adversaries may explicitly employ a known encryption algorithm to conceal command and control traffic rather than relying on any inherent protections provided by a communication protocol. Despite the use of a secure algorithm, these implementations may be vulnerable to reverse engineering, if necessary, secret keys are encoded and/or generated within malware samples/configuration files.
	T1474	Supply chain compromise	As further described in Supply Chain Compromise, supply chain compromise is the manipulation of products or product delivery mechanisms prior to receipt by a final consumer for the purpose of data or system compromise. Somewhat related, adversaries could also identify and exploit inadvertently present vulnerabilities. In many cases, it may be difficult to be certain whether exploitable functionality is due to malicious intent or simply inadvertent mistake.
	T1508	Suppress application icon	A malicious application could suppress its icon from being displayed to the user in the application launcher to hide the fact that it is installed, and to make it more difficult for the user to uninstall the application. Hiding the application's icon programmatically does not require any special permissions.
	T1426	System information discovery	An adversary may attempt to get detailed information about the operating system and hardware, including version, patches, and architecture.

(continued)

Table 5 (continued)

Enterprise			Mobile		
ID	Name	Description	ID	Name	Description
			T1422	System network configuration discovery	On Android, details of onboard network interfaces are accessible to apps through the java.net.NetworkInterface class. The Android TelephonyManager class can be used to gather related information such as the IMSI, IMEI, and phone number.
			T1421	System network connections discovery	On Android, applications can use standard APIs to gather a list of network connections to and from the device. For example, the Network Connections app available in the Google Play Store advertises this functionality.
			T1509	Uncommonly used port	Adversaries may use non-standard ports to exfiltrate information.
			T1576	Uninstall malicious application	Adversaries may include functionality in malware that uninstalls the malicious application from the device. This can be achieved by:
			T1416	URI hijacking	Adversaries may register Uniform Resource Identifiers (URIs) to intercept sensitive data.
			T1618	User evasion	Adversaries may attempt to avoid detection by hiding malicious behaviour from the user. By doing this, an adversary's modifications would most likely remain installed on the device for longer, allowing the adversary to continue to operate on that device.
			T1481	Web service	Adversaries may use an existing, legitimate external Web service as a means for relaying commands to a compromised system.

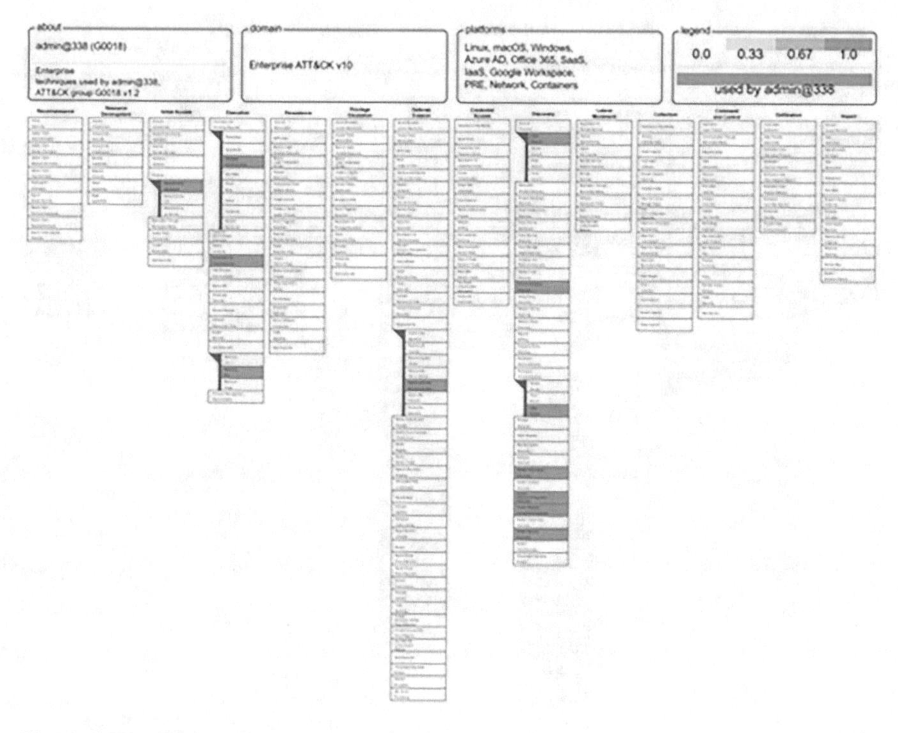

Fig. 6 G0018 Cyber kill chain

Additionally, the MITRE CAPEC framework can provide you with further threat insights into these known tactics, e.g. CAPEC-163: Spear Phishing (MITRE, 2022i).

Execution Flow

- **Explore**
  Obtain useful contextual detailed information about the targeted user or organization: An adversary collects useful contextual detailed information about the targeted user or organization in order to craft a more deceptive and enticing message to lure the target into responding.
- **Techniques**
  Conduct web searching research of target. See also: CAPEC-118.
  Identify trusted associates, colleagues and friends of target. See also: CAPEC-118.
  Utilize social engineering attack patterns such as Pretexting. See also: CAPEC-407.
  Collect social information via dumpster diving. See also: CAPEC-406.
  Collect social information via traditional sources. See also: CAPEC-118.
  Collect social information via Non-traditional sources. See also: CAPEC-118.
- **Experiment**
  Optional: Obtain domain name and certificate to spoof legitimate site: This optional step can be used to help the adversary impersonate the legitimate site more convincingly. The adversary can use homograph attacks to convince users that they are using the legitimate website. Note that this step is not required for phishing attacks, and many phishing attacks simply supply URLs containing an IP address and no SSL certificate.

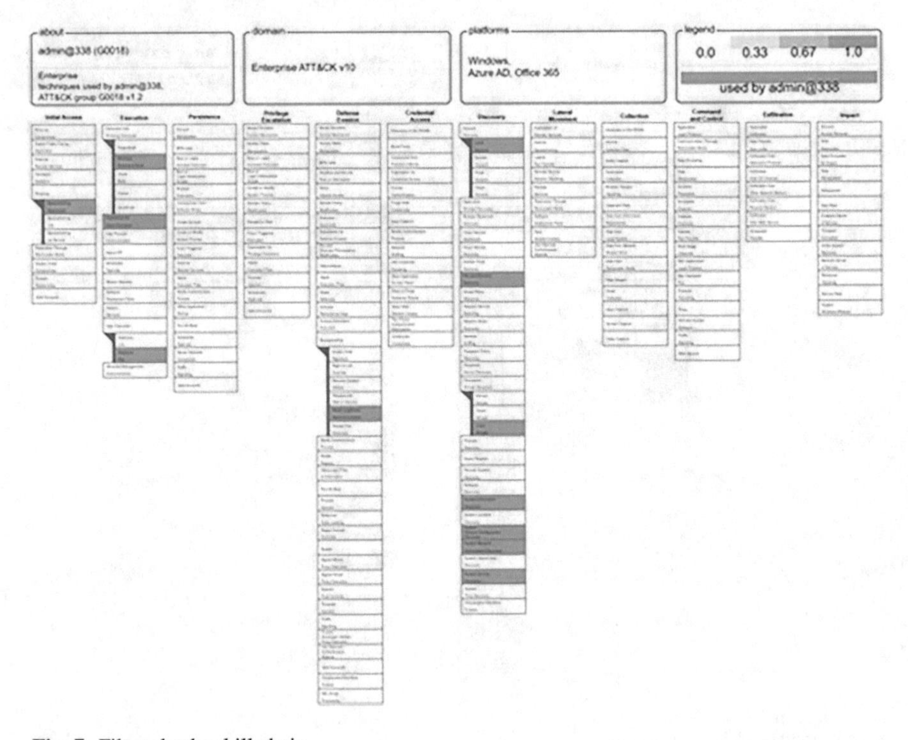

Fig. 7 Filtered cyber kill chain

- **Techniques**
  Optionally obtain a domain name that visually looks similar to the legitimate site's domain name. An example is www.paypaI.com vs. www.paypal.com (the first one contains a capital i, instead of a lower-case L).
  Optionally obtain a legitimate SSL certificate for the new domain name.
  Optional: Explore legitimate website and create duplicate: An adversary creates a website (optionally at a URL that looks similar to the original URL) that closely resembles the website that they are trying to impersonate. That website will typically have a login form for the victim to put in their authentication credentials. There can be different variations on a theme here.
- **Techniques**
  Use spidering software to get copy of web pages on legitimate site.
  Manually save copies of required web pages from legitimate site.
  Create new web pages that have the legitimate site's look at feel but contain completely new content.
  Optional: Build variants of the website with very specific user information e.g., living area, etc.: Once the adversary has their website which duplicates a legitimate website, they need to build very custom user related information in it. For example, they could create multiple variants of the website which would target different living area users by providing information such as local news, local weather, etc. so that the user believes this is a new feature from the website.

- **Techniques**
  Integrate localized information in the web pages created to duplicate the original website. Those localized information could be dynamically generated based on unique key or IP address of the future victim.
  Exploit
  Convince user to enter sensitive information on adversary's site.: An adversary sends a message (typically an e-mail) to the victim that has some sort of a call to action to get the user to click on the link included in the e-mail (which takes the victim to adversary's website) and log in. The key is to get the victim to believe that the message is coming from a legitimate entity trusted by the victim or with which the victim or does business and that the website pointed to by the URL in the e-mail is the legitimate website. A call to action will usually need to sound legitimate and urgent enough to prompt action from the user.
- **Techniques**
  Send the user a message from a spoofed legitimate-looking e-mail address that asks the user to click on the included link.
  Place phishing link in post to online forum.
  Use stolen credentials to log into legitimate site: Once the adversary captures some sensitive information through phishing (login credentials, credit card information, etc.) the adversary can leverage this information. For instance, the adversary can use the victim's login credentials to log into their bank account and transfer money to an account of their choice.
- **Techniques**
  Log in to the legitimate site using another user's supplied credentials.

7 Conclusion

In summary, if your business is not carrying out *(or dedicating sufficient resources to)* threat prediction and modelling and providing this output into your risk committees/teams, your risk management practices will be fundamentally flawed.

Effective threat prediction and modelling helps an organization to be 'Forewarned' so that they can build suitable defensive strategies ('Forearmed'), based upon informed risk decision-making to significantly reduce the opportunities that are presented to a threat actor and then, to supply the ability to substantially reduce the risks to a business' valued operations/assets.

Your threat prediction and modelling should be the drivers for effective defensive strategy development, whilst supplying more reassurances, to any risk committees and the C-Suite members, that any mitigation measures are aligned to address any identified threats.

References

Beattie, A. (2021, December 15). Corporate espionage: Fact and fiction. *Investopedia*. www.investopedia.com/financial-edge/0310/corporate-espionage-fact-and-fiction.aspx

Blank, R., & Gallagher, P. (2012). *Guide for conducting risk assessments NIST Special Publication 800-30 Revision 1 Joint Task Force Transformation Initiative*.

Cambridge Dictionary. (2019a). ASSET | Meaning in the Cambridge English Dictionary. Cambridge.org, dictionary.cambridge.org/dictionary/english/asset

Cambridge Dictionary. (2019b, September 25). THREAT | Meaning in the Cambridge English Dictionary. Cambridge.org, dictionary.cambridge.org/dictionary/english/threat

Cambridge Dictionary. VULNERABILITY | Meaning in the Cambridge English Dictionary. Dictionary.cambridge.org, dictionary.cambridge.org/dictionary/english/vulnerability

Clark, T. (2020, December 5). Threat model | Flow & attack tree diagram | Payment process. GitHub. github.com/TClark000/threat-models. Accessed 22 Feb 2022.

Colley, M. P., & Development, Concepts and Doctrine Centre. (2011). *Understanding and intelligence support to joint operations* (pp. 2–16). Development, Concepts and Doctrine Centre.

CPNI. (2014). Terrorism is a major threat for businesses | Public Website. Cpni.gov.uk, www.cpni.gov.uk/terrorism

CPNI. (2020). The UK is a high priority espionage target | Public Website. Cpni.gov.uk, www.cpni.gov.uk/espionage

CSRC. (2015). Vulnerability – Glossary | CSRC. Nist.gov, csrc.nist.gov/glossary/term/vulnerability

Cybersecurity & Infrastructure Security Agency. (2022, January 11). Understanding and mitigating Russian state-sponsored cyber threats to U.S. Critical Infrastructure | CISA. www.cisa.gov, www.cisa.gov/uscert/ncas/alerts/aa22-011a

de León, R. (2020, December 17). 50% of U.S. tech execs say state-sponsored cyber warfare their biggest threat: CNBC Survey. *CNBC*. www.cnbc.com/2020/12/17/50percent-of-tech-execs-say-cyber-warfare-biggest-threat-cnbc-survey.html. Accessed 12 Feb 2022

Dempsey, K., et al. (2017, June). *Automation support for security control assessments. Vol. 1: Overview*. nvlpubs.nist.gov/nistpubs/ir/2017/NIST.IR.8011-1.pdf, https://doi.org/10.6028/nist.ir.8011-1. Accessed 29 Oct 2020.

Dictionary.com. (2022). Definition of sabotage | Dictionary.com. www.dictionary.com, www.dictionary.com/browse/sabotage. Accessed 12 Feb 2022.

EC Council. Threat modeling | Importance of threat modeling. EC-Council. www.eccouncil.org/threat-modeling

Editor, CSRC Content. Threat scenario – Glossary | CSRC. Csrc.nist.gov, csrc.nist.gov/glossary/term/threat_scenario

Ejoh, E. (2021, December 28). Pipeline vandalism, sabotage, others culminate in shutdown of 8 oil terminals – report. *Vanguard News*. www.vanguardngr.com/2021/12/pipeline-vandalism-sabotage-others-culminate-in-shutdown-of-8-oil-terminals-report. Accessed 12 Feb 2022.

Etymology Online Dictionary. (2022a). Threat | Search online Etymology Dictionary. www.etymonline.com, www.etymonline.com/search?q=threat. Accessed 11 Feb 2022.

Etymology Online Dictionary. (2022b). Vulnerable | Etymology, origin and meaning of vulnerable by Etymonline. www.etymonline.com, www.etymonline.com/word/vulnerable#etymonline_v_7899. Accessed 11 Feb 2022.

GeeksforGeeks. (2021, November 8). Threat modelling. GeeksforGeeks. www.geeksforgeeks.org/threat-modelling. Accessed 22 Feb 2022.

Hell, M. (2021, July 29). What is a security threat? *Debricked*. debricked.com/blog/what-is-security-threat. Accessed 11 Feb 2022.

ISACA. (2013). *Advanced persistent threats: How to manage the risk to your business*. ISACA.

jegeib. (2022, January 3). Threats – Microsoft Threat Modeling Tool – Azure. Docs.microsoft.com, docs.microsoft.com/en-us/azure/security/develop/threat-modeling-tool-threats

Justia Patents. (2021). Patents assigned to Croda, Inc. – Justia Patents Search. Patents.justia.com, patents.justia.com/assignee/croda-inc. Accessed 12 Feb 2022.

Kim, Y.-G., & Cha, S. (2011, April 7). Threat scenario-based security risk analysis using use case modeling in information systems. *Security and Communication Networks, 5*(3), 293–300. https://doi.org/10.1002/sec.321

Luo, F., et al. (2021, September 21). Threat analysis and risk assessment for connected vehicles: A survey. *Security and Communication Networks*, *2021*, 1–19. https://doi.org/10.1155/2021/1263820. Accessed 11 Oct 2021.

Matilainen, J. (2021). *Using cyber threat intelligence as a part of organisational cybersecurity*. Jyx.jyu.fi. urn.fi/URN:NBN:fi:jyu-202105313334.

MITRE. (2022a). Tactics – Enterprise | MITRE ATT&CK®. Attack.mitre.org, attack.mitre.org/tactics/enterprise. Accessed 22 Feb 2022.

MITRE. (2022b). Tactics – Mobile | MITRE ATT&CK®. Attack.mitre.org, attack.mitre.org/tactics/mobile. Accessed 22 Feb 2022.

MITRE. (2022c). Mitigations – Enterprise | MITRE ATT&CK®. Attack.mitre.org, attack.mitre.org/mitigations/enterprise. Accessed 22 Feb 2022.

MITRE. (2022d). Mitigations – Mobile | MITRE ATT&CK®. Attack.mitre.org, attack.mitre.org/mitigations/mobile. Accessed 22 Feb 2022.

MITRE. (2022e). Account Discovery, Technique T1087 – Enterprise | MITRE ATT&CK®. Attack.mitre.org, attack.mitre.org/techniques/T1087. Accessed 22 Feb 2022.

MITRE. (2022f). Operating System Configuration, Mitigation M1028 – Enterprise | MITRE ATT&CK®. Attack.mitre.org, attack.mitre.org/mitigations/M1028. Accessed 22 Feb 2022.

MITRE. (2022g). Admin@338, Group G0018 | MITRE ATT&CK®. Attack.mitre.org, attack.mitre.org/groups/G0018. Accessed 22 Feb 2022.

MITRE. (2022h, February 22). ATT&CK® Navigator. Mitre-Attack.github.io, mitre-attack.github.io/attack-navigator//#layerURL=https%3A%2F%2Fattack.mitre.org%2Fgroups%2FG0018%2FG0018-enterprise-layer.json. Accessed 22 Feb 2022.

MITRE. (2022i). CAPEC – CAPEC-163: Spear Phishing (Version 3.6). Capec.mitre.org, capec.mitre.org/data/definitions/163.html. Accessed 22 Feb 2022.

MITRE. CAPEC – Common Attack Pattern Enumeration and Classification (CAPECTM). Capec.mitre.org, capec.mitre.org/index.html

MITRE. Groups | MITRE ATT&CK®. Attack.mitre.org, attack.mitre.org/groups

MITRE. MITRE ATT&CK®. Attack.mitre.org, attack.mitre.org

National Initiative for Cybersecurity Careers and Studies (NICCS). Cybersecurity Glossary | National Initiative for Cybersecurity Careers and Studies. Niccs.cisa.gov, niccs.cisa.gov/about-niccs/cybersecurity-glossary

NATO Standardization Office (NSO). (2015). NATO Standard AJP-3.14 Allied Joint Doctrine for Force Protection Edition a Version 1 with UK National Elements.

NCSC. (2017, April 9). *NCSC publishes new report on criminal online activity*. www.ncsc.gov.uk, www.ncsc.gov.uk/news/ncsc-publishes-new-report-criminal-online-activity

NIST. (2015). Threat – Glossary | CSRC. Nist.gov, csrc.nist.gov/glossary/term/threat

Online Etymology Dictionary. (2022). Assets | Etymology, origin and meaning of assets by Etymonline. www.etymonline.com, www.etymonline.com/word/assets#etymonline_v_17964. Accessed 11 Feb 2022.

Questys Solutions. (2022). I2 intelligence analysis and advanced analytics platform | I2. I2group.com, i2group.com. Accessed 12 Feb 2022.

Ransomware Database. (2022, February 11). *Ransom-DB – Ransomware Groups*. www.ransom-Db.com, www.ransom-db.com/ransomware-groups. Accessed 11 Feb 2022.

Ross, R., et al. (2021, December). *NIST Special Publication 800–160. Developing cyber-resilient systems: A systems security engineering approach* (Vol. 2). nvlpubs.nist.gov/nistpubs/SpecialPublications/NIST.SP.800-160v2r1.pdf, https://doi.org/10.6028/NIST.SP.800-160v2r1. Accessed 30 Jan 2022.

Saini, V. K., et al. (2008, April). *Threat modeling using attack trees*. www.researchgate.net/publication/234738557_Threat_Modeling_Using_Attack_Trees. Accessed 22 Feb 2022.

Shetty, V. (2018, January 13). How does weather forecasting work?. *Science ABC*. www.scienceabc.com/innovation/how-does-weather-forecasting-work.html

Shirey, R. (2007, August). Rfc4949. Datatracker.ietf.org, datatracker.ietf.org/doc/html/rfc4949

Shirey, R. (2013, March 2). RFC 2828 – Internet security glossary. Sandbox-Cf.ietf.org, sandbox-cf.ietf.org/doc/rfc2828. Accessed 11 Feb 2022.

Soetrust. (2021, October 1). Definition of subversion – What it is, meaning and concept – Soetrust. *Soetrust.* soetrust.org/misc/definition-of-subversion-what-it-is-meaning-and-concept. Accessed 12 Feb 2022.

Spencer, C. (2021, July 30). FBI investigating 41 cases of eco-sabotage in Washington. *TheHill.* thehill.com/changing-america/sustainability/environment/565729-fbi-investigating-41-cases-of-eco-sabotage-in. Accessed 12 Feb 2022.

Statistica. (2021, February 16). Leading cause of ransomware infection 2020. *Statista.* www.statista.com/statistics/700965/leading-cause-of-ransomware-infection. Accessed 11 Feb 2022.

Stine, K., et al. (2020, October 13). *Integrating Cybersecurity and Enterprise Risk Management (ERM).* NISTIR 8286 Integrating Cybersecurity and Enterprise Risk Management (ERM). nvlpubs.nist.gov/nistpubs/ir/2020/NIST.IR.8286.pdf, https://doi.org/10.6028/nist.ir.8286

The additional ‘SS’ has been added to represent global government’s use of cyber criminals against their enemies.

Tounsi, W., & Rais, H. (2018). A survey on technical threat intelligence in the age of sophisticated cyber attacks. *Computers & Security, 72*, 212–233. https://doi.org/10.1016/j.cose.2017.09.001

trikers. (2022). Trike. Octotrike.org, www.octotrike.org. Accessed 22 Feb 2022.

Tzu, S. (2019). *The art of war* (p. 3:6:1–6). Ixia Press.

Wadhwani, S. (2022). *Super malicious insiders responsible for a third of all insider security incidents*. Toolbox.com. www.toolbox.com/it-security/threat-reports/news/insider-risk-insider-threat-rising-super-malicious-insider. Accessed 12 Feb 2022.

Welekwe, A. (2021, September 13). *Threat modeling guide*. Comparitech. www.comparitech.com/net-admin/threat-modeling-guide. Accessed 22 Feb 2022.

Weston, S. (2021, May 26). *Threat modeling 101: Getting started with application security threat modeling [2021 Update]*. Infosec Resources. resources.infosecinstitute.com/topic/applications-threat-modeling. Accessed 22 Feb 2022.

Wunder, J., et al. (2011, June). *Specification for asset identification 1.1*.

A Critical Analysis of the Dark Web Challenges to Digital Policing

Ellie Moggridge and Reza Montasari

Abstract This chapter critically analyses the challenges to the digital policing caused by the anonymity and growing criminal market of the Dark Web. The police face many ethical and jurisdictional issues when monitoring illicit activity, particularly in the case of online paedophilia and child pornography that circulates the web. Issues arise of who is responsible for prosecuting crimes committed on international scales and how internet crimes can be detected in an ethical way that maintains a degree of privacy and privacy rights to its legitimate users. Enforcement agencies have adapted effective investigative techniques such as NIT to takedown illicit sites such as 'Playpen' and have demonstrated a degree of success. The current strengths of digital policing can be recognised, yet international cooperation and communications within the private sector demonstrate further preventative measures.

Keywords Dark Web · Digital policing · NIT · Child pornography · Anonymity · Police challenges · Investigative techniques · Deep web · Surface web · Honeypot

1 Introduction

This chapter aims to address the challenges faced in the world of digital policing, and more specifically the challenges caused by the emergence of the Dark Web. It is important to address issues as such due to the accessibility of the internet that now enables the practice of illegitimate activities to often go undetected. Historically, the Dark Web may have appeared to be a space beyond reach of law enforcement with its technological anonymity protecting criminals from detection, arguably, this

E. Moggridge (✉) · R. Montasari
School of Social Sciences, Department of Criminology, Sociology and Social Policy, Swansea University, Swansea, UK
e-mail: 852895@Swansea.ac.uk; Reza.Montasari@Swansea.ac.uk
URL: http://www.swansea.ac.uk

R. Montasari (ed.), *Artificial Intelligence and National Security*,
https://doi.org/10.1007/978-3-031-06709-9_8

is no longer becoming the case, with law enforcement now deploying an array of techniques to identify and convict perpetrators (Davies, 2020). Traditional policing techniques have required amendment and development to break the Dark Web's use of encryption. It is therefore important to note that within this chapter, the main objective will be to critically identify and evaluate the issues and challenges posed to digital policing and later offer justified solutions.

The arising problems will be well researched and evidenced in digital police practice to provide an in-depth insight into the problems that the Dark Web poses. To achieve the research aim, existing articles, journals and chapters will be surveyed and critically analysed. Subsequently, this chapter will begin with background research into the Dark Web and digital policing. This section provides a broad overview into what the Dark Web is, its history, how it is accessed, and distinguishes the different layers that encompass the internet. This section works to provide a contextualised account of digital policing and the Dark Web. Following the overview, comes the bulk of the chapter which will provide a critical insight into the issues and its implications on digital policing. The following section will aim to propose effective solutions to the consequences discussed previously that can be put in place to effectively target the issues associated with digital policing and the Dark Web. Significant key findings will be contextualised and drawn with supporting evidence and criticisms in the form of a discussion. Finally, this chapter will conclude with an overall review of the research problem, highlighting the main findings of this chapter and its contributions to already existing work in this field.

2 Background

To address the issues derived from the Dark Web, it is important to first determine the concept of cyberspace and the elements it comprises of. The internet and World Wide Web are commonly used interchangeably; however, the two terms mean separate things (Global Commission on Internet Governance Chapter Series, 2015). Whilst the internet connects users globally, forming 'communications with any other computer', the World Wide Web is an information-sharing model built on top of the internet. In 2014, internet users reached a magnitude of over 3 billion surfers worldwide; however, whilst there is research into its population, there is little data that provides insight into the numbers of those users that access the varying layers that make up the web (Finklea, 2017). '*The internet comprises every single server, computer and other device that is connected together in a network of networks*' which is then divided into two elements, being the surface web and the deep web. (Chertoff, 2017).

The deep web is characterised by the 'unknown breadth, depth, content and users' (Finklea, 2017; 1). Moreover, the term refers to '*a class of content on the Internet that, for various technical reasons, is not indexed by search engines,' and 'thus would not be accessible through a traditional search engine*' (Finklea, 2017; 1). The reasoning behind this may be intentional or may be due to the nature of the website itself (Zhang & Zou, 2020). Owners of such websites can restrict this access through a number of ways, for example '*restricting access to pages through technical means*' and by ensuring no open '*websites contain links to their pages*' (Zhang & Zou, 2020; 1694). The surface web on the other hand is what an average user would consider as 'the internet', whereby it consists of a collection of websites 'indexed by search engines such as Google, Yahoo and Bing', and this can be accessed easily through a standard browser (Chertoff, 2017). It is used routinely and consists of data that search engines can find and then offer up in response to queries and is commonly connoted as being just the 'tip' of the iceberg (Global Commission on Internet Governance Chapter Series, 2015). This is because the traditional search engine supposedly only sees about 0.03% of the information available, with the rest of its information 'submerged' into the deep web (Bergman, 2001; 1). The Dark Web is a proportion of the deep web in which has been made intentionally inaccessible through a standard web browsing (Chertoff & Simon, 2015). To fully understand the scale of the Dark Web, it is important to note that researchers predict it is between 4000 and 5000 times larger than that of the surface web and accounts for around 90% of the traffic on the internet (Chertoff, 2017). The Dark Web refers to an encrypted network that can only be accessed using dedicated software in which will later be discussed. These forms of networks rely on routing traffic through the encryption layer to support anonymity within its user.

In the late 1990s, two research organisations in the US department of Defence developed an '*anonymised and encrypted network that would protect the sensitive communications of US spies*' and this network would not be inaccessible to ordinary internet surfers. (Kumar & Rosenbach, 2019; 22). The browser 'TOR' plays an essential role in accessing the Dark Web, originating as 'The onion router' with its many layers of encryption that guard passing information. Its main purpose was to separate identification information from routing and to design an anonymous communication network for military communication (Jadoon et al., 2019). The browser consists of a '*global overlay networks of relays which helps in the achievement of privacy and anonymity for user Internet traffic*' (Jadoon et al., 2019; 60). It consists of more than 7000 relay stations to redirect internet traffic and in result conceal its users' identity, making it difficult to track activity (Zhang & Zou, 2020). Whilst anonymity allows for privacy and promotes freedom, it can also be abused for illegal activities (Zhang & Zou, 2020).

The Dark Web facilitates a growing marketplace in which criminals can use to traffic drugs, stolen identities, child pornography, and various other illicit activities. Additionally, they can be purchased through an untraceable cryptocurrency which requires '*close cooperation between law enforcement, financial institutions, and regulators around the world is required to tighten the screws on nefarious activity*' (Kumar & Rosenbach, 2019; 22). TOR, however, has not been the only development that enabled the creation of the Dark Web, instead, 'Hidden Wiki' and 'Bitcoin' serve additional practical purposes that enable the Dark Web. Bitcoin became the standard currency of the Dark Web, encrypted in digital wallets and designed to be difficult to track back to the person who has spent them. Emerging in 2011, Bitcoin has become the currency of choice for illicit activities on the Dark Web such as the Silk Road (Kumar & Rosenbach, 2019).

3 Challenges

Studies revealed that child pornography generates the most traffic to the hidden sites on 'TOR', with sites such as 'Playpen' generating over 200.000 members distributing photos and videos of child pornography (Kaur & Randhawa, 2020). Figures as such emphasise the need for law enforcement to develop effective strategies to detect these acts and secure convictions. A common strategy incorporated in digital policing is the use of undercover investigations, whereby police infiltrate online forums in order to obtain crucial evidence and secure a conviction. In order to do so, police must establish a 'controlled surveillance operation' in which officers impersonate an administrator or moderator of an illicit forum (Davies, 2020). The priorities of such operations include, investigating chat rooms, newgroups, and peer-to-peer networks (Davies, 2020). Task Force Argos is an example of an undercover investigation, established in the late 1990s, the task force was the first branch of the Queensland Police service that set out to address the issues of child exploitation material. Whilst undercover, task Force Argos were successful in identifying a number of repeat offenders. Moreover, the covert operation Artemis of 2013 which resulted in the closure of 'Child's play' lead to a further 90 primary targets worldwide and 900 users being identified for exploitation material (Bleakley, 2018). Further success is demonstrated in the United States, whereby a LEA covertly took over the account of a staff member on the 'Silk Road' and continued to operate undercover and gather insight into illicit activities (Davies, 2020). However successful these strategies may appear, Wells et al. (2007) highlight the difficulties that undercover investigations pose at securing a conviction (Fig. 1).

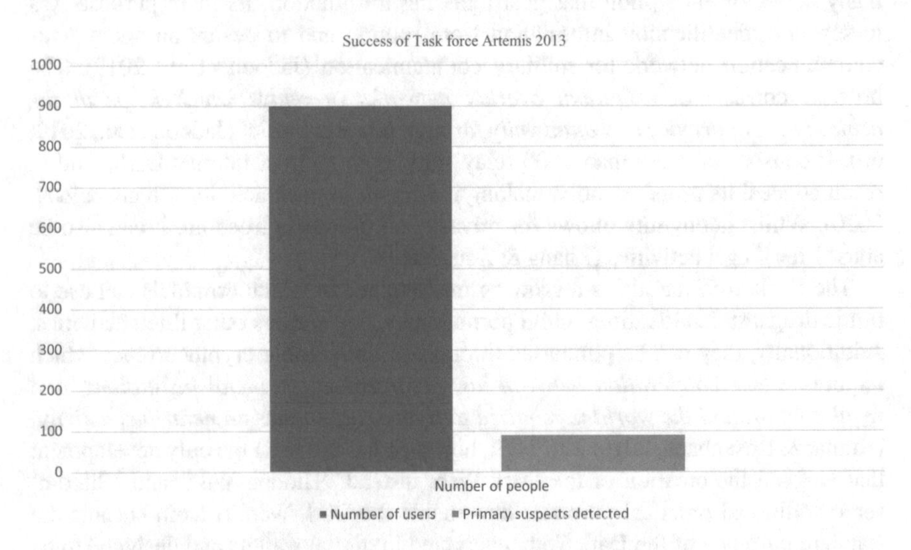

Fig. 1 Data illustrating the success of the undercover taskforce Artemis

It is argued covert investigations are not enough to charge an individual with possession, as the victim may be difficult to be identified and proven to be under a legal age of consent (Bleakley, 2018.) Instead there calls for a 'globally-coordinated approach' as an alternative approach to tackling the issue (Wells et al., 2007). A further criticism is that due to the nature of the internet itself, there is no single controlling agency as the child pornography may have been produced in one country yet accessed in another. The implications being that now there are multiple law enforcement agencies involved and the original location of the image is unknown creating ambiguity as to what jurisdiction is responsible for the prosecution (Haaszf, 2016). A further case that incorporated covert undercover investigating into police practice was 'Operation Pacifier' an FBI lead investigation that resulted in the shutdown of notorious TOR based website 'Playpen' whereby in the 2 weeks that the FBI operated the site, Network investigative techniques (NITs) were used to hack the computers of those that visited the sites and 900 arrests were made (Leukfeldt & Holt, 2020). The operation, however, has been criticised by alleged Playpen users on the validity of the NIT warrant. The case of Yang Kim in New York District Court raised the issue that the use of NITs to gather evidence for his possession of child pornography constituted a breach of his Fourth Amendment constitutional right to not be subject to unreasonable search and seizures (Leukfeldt & Holt, 2020). Most importantly, there is a great deal of controversy around the ethics and morality of undercover policing such a sensitive topic. Furthermore, a Dutch government chapter concluded that '*Moral and societal acceptability can also apply as an objection; the government should not be involved in such crimes, not even through an undercover agent*' (Vendius, 2015; 13). Additionally, it can be argued that this strategy contributes to the revictimisation of child sexual abuse victims. Varying academic opinions suggest there is an ethical grey area regarding this method of policing, arguing the practice of authorised criminality is secret, unaccountable, and in conflict with some of the basic premises of 'democratic policing' (Joh, 2009; 157). On the one hand, it can be argued that '*engaging in criminal behaviour to gather evidence is ordinarily considered inappropriate, there is also a widespread agreement that these tactics work when it comes to infiltrating online child exploitation networks*' (Bleakley, 2018; 224).

Another method used by police to enforce law on the Dark Web is the use of 'honeypot traps'. If criminals themselves have become more advanced criminals through technology, then law enforcement will need to be more efficient at targeting crime through technology, through the use of honeypot traps, which are '*constitutionally permissible and a proactive means to combat child exploitation on the dark web*' (Gregory, 2018; 261). The strategy behind the use of honeypots is to '*shut intruders safely from production systems for convert analysis and to obtain intelligent on the intruder by monitoring and logging every actions the intruder makes including access attempts, and files accessed*' (Rong & Yang, 2003; 186). There are similar concerns with the use of honeypot traps, in that they may be seen as a form of violation of civil liberties, this is because the purpose of the traps is to create '*uncertainty in the minds of offenders and reduce the sense of freedom and anonymity*' (Davis, 2017; 1). Furthermore, data indicates that honeypot traps are not the most effective strategy to police the Dark Web and therefore there may be other, more appropriate solutions (Davis, 2017).

Hacking is also a common method of digital policing in order to tackle criminal activity on the Dark Web. These hacking techniques use surveillance software to directly exert control of criminals and access devices. (Davis, 2017). Used effectively, this strategy can '*force the target computer to covertly upload files to a server controlled by law enforcement or commandeer computers that associate with the target by, for example, accessing a website it hosts*' (Davis, 2017; 1). The most effective way of determining who is behind an online crime is to trick the target into downloading malicious code – more specifically, what is referred to as a 'Network investigative technique or NIT' (Kerr & Murphy, 2017). NIT searches the target computer for location information and identity and sends this information to the government (Kerr & Murphy, 2017). A case that whereby these techniques were deployed was the February 2015 NIT warrant, whereby U.S Magistrate Judge Theresa Carrol Buchanan issued the search warrant of any person's computer that accessed the 'Playpen' site (Aucoin, 2018). The NIT warrant resulted in 8000 computers hacked by the FBI, obtaining the TOR users masked IP address to obtain a subpoena from the associated ISP. However, this case has been undermined under Rule 41 of the Federal Rules of Criminal Procedure and that the NIT warrant fails to meet the Fourth Amendment's requirements of probable cause and particularity (Aucoin, 2018).

The Federal Rule of Criminal procedure 41 permits a magistrate to issue search warrants for a device's location if the location has '*been concealed through technological means*' (Davis, 2017). The implication of this is that it may result in the largest expansion of extraterritorial enforcement jurisdiction in FBI history, with figures suggesting that approximately 80% of the computers on the Dark Web located outside the United States, it is likely that '*any given law enforcement target is likely to be located abroad which will pose sovereignty challenges between law enforcement agencies*' (Davis, 2017; 1). Moreover, the use of hacking can be difficult to incorporate in policing as is its use of NITs to investigate users of anonymising software presents a '*looming flashpoint between criminal procedure and international law*' as actions are argued to '*violate the sovereignty of other nations*' (Kerr, 2018; 59). This raises foreign relation risks due to the fact that it is unknown where the computers searched are located and in result, it may offend 'foreign nations and violate customary law's prohibition on a state's extraterritorial exercise of law enforcement functions without consent' and may lead to criminal prosecution of US officials abroad. On the other hand, the idea that this technique poses 'international relations risk' has been criticised on the understanding that this view instead overlooks the pervasive nature of transnational law enforcement cooperation in Dark Web investigations. Moreover, there is supporting evidence such as the emergence of bilateral treaties in which address cybercrime with a focus of mutual legal assistance (Kerr, 2018). Instead the extraterritorial aspects of network investigative techniques demonstrate a need for '*new substantive and procedural regulations that balance law enforcement goals with countervailing foreign relations interests*' (Ghappour, 2016; 1086).

The Dark Web is not always used illegally, instead the military, police, journalists and whistle-blowers may operate the browser in order to chapter wrongdoing within their employment without the risk of retaliation and with the ability to do so anonymously (Schein & Trautman, 2020). The TOR browser helps specific groups such as human rights activists, civil activists, journalists and whistle-blowers voice their opinion without the fear of being traced, and further supports the intent behind the Bill of Rights and the First Amendment; to protect unpopular voices persons from retribution by an 'intolerant society' (Priyadarshan, 2022). It is for these reasons that policing the Dark Web becomes difficult; law enforcement must tailor their policy so that they strike a balance between the needs of privacy-minded users and the government's responsibility to detect and punish illegal activity as previously discussed (Chertoff, 2017). The United States for example is constitutionally committed to protecting freedom of speech on the internet and so policing the anonymous browsers pose challenges and raise the issues of the arbitrary 'balancing act' – in which consists of the act of deciding when the public disclosure outweighs the interest of the organisation'(Webber, 1995).

4 Solutions

The internet by nature is an international network of computers, which indicates that an international approach is required in order to work towards an effective solution to crime on the Dark Net. The main agenda in terms of tackling this issue, lies with regulating 'TOR' and in order to do so, there must be solidified policies on how to regulate this within a legal framework; however, in order for solutions to be long lasting, policies must be clear and 'internationally agreeable' (Chertoff, 2017; 32). Nonetheless, international collaboration poses a challenge, to work internationally without compromising each countries individual ideals. An example of this challenge being that some members of the international community that have pushed differing agendas related to TOR that are inconsistent with the aims of the USA (Chertoff, 2017). Yet in the recent decades, there has been an increase in international cooperation and investigation into cybercrime; with embassies around the world that are now equipped with the appropriate measures to investigate transnational crimes and communicate nationally, through transnational phone calls, faxes, emails and internet searches (Kerr & Murphy, 2017). The case of 'Darkode' illustrates further success of international cooperation, being described '*a coordinated effort by a coalition of law enforcement authorities from 20 nations to charge, arrest or search 70 Darkode members and associates around the world*' and demonstrated the '*largest coordinated international law enforcement effort ever directed at an online cyber-criminal forum*' (Schein & Trautman, 2020).

Additionally, recent Justice Department chapters represented this importance to pool resources among all participating countries which lead to more than 200 TOR operated child pornography sites being taken down. This case emphasised the importance for international communication, so that a logical and effective Dark

Web policy can be established and practiced to govern the Dark Web. It is vital that leading law enforcement agencies facilitate '*more dialogues on trans-border cybercrime offending*' through international conferences, as they enable informal cooperation and in result reduce the likelihood of '*bureaucratic entanglements*'; and delays in obtaining evidence electronically, which prove vital to securing a conviction. (Brown, 2015; 90). There have been criticisms of cross border cooperation being internationally wrong with law enforcement investigative techniques undermining the 'sovereignty of nations' (Ghappour, 2016). Yet supporting evidence draws on the norms of international cooperation rather than confrontation and undermines this risk of deteriorating international relations (Kerr & Murphy, 2017). In fact, it can be argued that the growing threat of 'TOR' is actually increasing government cooperation rather than raise '*fears of territorial encroachment*' (Kerr & Murphy, 2017; 62).

In addition to collaborating on an international scale, it is important to collaborate and communicate effectively with the private enterprises that develop the necessary technology. This is crucial due to the fact that large-scale data breaches are likely to target private companies that carry personal information such as healthcare (Romeo, 2016). An example of this coordination in action is the breakup of a large botnet by European Law enforcement and Microsoft. Furthermore, the introduction of The Consumer Privacy Acts of 2015 in which imposed a degree of duty on private companies in the event of a data breach (Romeo, 2016). Communication will work to strengthen private sector infrastructure and employs preventative measures to tackle the issue (Romeo, 2016). Whilst law enforcement has the '*jurisdiction to pursue criminals*', it is important to note that private companies own and operate much of the software and hardware of the internet, indicating a need for public-private partnerships between law enforcement and these companies. If communication is effective, this collaboration will act as a force multiplier, increasing the efficiency of online policing (Jardine, 2015; 47). Local and national collaboration is vital however, it is important to specialise skills so that efforts are the most efficient; for example, local police should not attempt to bust 'international online frauds' and instead each level should 'stick to its strengths' (Jardine, 2015; 10j).

Even with international and public influence, it is clear that more training is required in order to fully equip law enforcement agencies with the adequate resources to tackle the threat of online crimes. Law enforcement agencies need to invest in cybercrime training for officers in order to effectively deal with the issue at hand. Many suggest local law enforcement are undertrained and underresourced when dealing with cybercrime. In particular, executive director of the Major City Chiefs Police Association states '*Most local police do not have the capacity to investigate these cases even if they have jurisdiction*' (Jardine, 2015). Agencies will need to train officers to '*safeguard the evidentiary integrity of a digital crime scene*' to ensure valuable evidence is not lost (Leukfeldt & Holt, 2020; 368). Nevertheless, the UK has aims of launching dedicated cybercrime units in every police force, with national training programmes for police in which are sponsored by the National Police Chiefs Council (Davies, 2020). Additional units such as JOC

(Joint Operations Cell) also focus on child sexual exploitation (Davies, 2020). This demonstrates the beginning of a smarter, more equipped taskforce with the adequate skills to tackle crimes on the Dark Web.

5 Discussion

The chapter highlights the issues of the Dark web and academia; clearly demonstrates the issues law enforcement face when tracking criminal activity on anonymous browsers such as TOR. Researchers suggest that illicit content accessible through TOR were clearly seen on 1547 of 2723 live sites, with most common themes of drugs, illicit finance and child and animal pornography (Davis, 2017). In terms of the issues faced, the chapter highlights the prevalence and growing market for child pornography on sites such as 'Playpen' in which the FBI estimates consists of around 215,000 users (Finklea, 2017) Supporting research portrays a general consensus for prioritising and tackling the issue of child exploitation material circulating the Dark Net. There have been many successes of police intervention through undercover investigations and 'honeypot' traps, which to an extent support the need to continue these forms of online policing; however, the reality is that empirical research or lack of it, suggests that actually there is little understanding of the network structure of these perpetrators and even less is known about the true impact of the police interventions (da Cunha et al., 2020). It is suggested that more research into the hidden figures and true prevalence of crime on the Dark Web is required to truly install effective police practice. More specifically, specialised crime units typically only address internet crimes that are against children, and whilst this is important, there is little focus on the wider issue such as fraud, identity theft and intrusion (Leukfeldt & Holt, 2020; 370). There has been a trend in academic research that highlights the success and need for international cooperation and communication of law enforcement policing techniques with further research highlighting the successes of organisations such as Darkode. (Kerr & Murphy, 2017). Prior issues of international relations difficulties have been found unpersuasive and are diminished by the overwhelming studies that demonstrate welcoming international cooperation and in result success in digital policing.

6 Conclusion

To conclude, this chapter raises the key issues surrounding digital policing on the dark net with its lasting implications, and provides preventative measures that may minimise the crimes committed on the Dark Web. The sophistication of the internet and the growing expertise of online criminals create challenges with Law enforcement agencies in many ways of which have been previously discussed.

Law enforcement agencies must adopt a balanced approach by which focusses on the exposure and prosecution of criminals, whilst refraining from interfering with '*innocuous activities and exercise political freedoms*' (Hadjimatheou, 2022). The reality is that online crime is thriving and growing on the internet due to the ability to now maintain anonymity and therefore evade law enforcement (Davis, 2017). Whilst it may seem the only true solution is to get rid of the Dark Web itself, this would not be feasible and instead law enforcement must adapt logical steps that strike a balance between protecting user anonymity and preventing illegal activities online (Chertoff, 2017). Critically examining past mistakes and problems on digital policing will allow for a more successful future of digital policing.

References

Aucoin, K. E. (2018). The Spider's Parlor: Government malware on the dark web. *Hastings Law Journal, 69*(5), 1433–1469.

Bergman, M. (2001). White chapter: The deep web: Surfacing hidden value. *The Journal of Electronic Publishing, 7*(1). https://doi.org/10.3998/3336451.0007.104

Bleakley, P. (2018). Watching the watchers: Taskforce Argos and the evidentiary issues involved with infiltrating dark web child exploitation networks. *The Police Journal: Theory, Practice and Principles, 92*(3), 221–236. https://doi.org/10.1177/0032258x18801409

Brown, C. S. (2015). Investigating and prosecuting cyber crime: Forensic dependencies and barriers to justice. *International Journal of Cyber Criminology, 9*(1), 55.

Chertoff, M. (2017). A public policy perspective of the dark web. *Journal of Cyber Policy, 2*(1), 26–38. https://doi.org/10.1080/23738871.2017.1298643

Chertoff, M., & Simon, T. (2015). *The impact of the dark web on internet governance and cyber security. Global Commission on Internet Governance, Chapter Series: No. 6*. Canadian Electronic Library/CIGI.

da Cunha, B. R., MacCarron, P., Passold, J. F., et al. (2020). Assessing police topological efficiency in a major sting operation on the dark web. *Scientific Reports, 10*, 73. https://doi.org/10.1038/s41598-019-56704-4

Davies, G. (2020). Shining a light on policing of the dark web: An analysis of UK investigatory powers. *The Journal of Criminal Law, 84*(5), 407–426. https://doi.org/10.1177/0022018320952557

Davis, C. (2017). *Addressing the challenges of enforcing the law on the dark web [Blog]*. Retrieved 23 December 2021, from https://law.utah.edu/addressing-the-challenges-of-enforcing-the-law-on-the-dark-web/

Finklea, K. (2017, March 10). *Dark web*. Congressional Research Service, 1–19. https://fas.org/sgp/crs/misc/R44101.pdf

Ghappour, A. (2016). Searching places unknown: Law enforcement jurisdiction on the dark web. *SSRN Electronic Journal, 69*(4), 1075–1136. https://doi.org/10.2139/ssrn.2742706

Global Commission on Internet Governance Chapter Series. (2015). *The impact of the dark web on internet governance and cyber security*. Centre for International Governance Innovation and Chatham House. Retrieved from https://www.cigionline.org/static/documents/gcig_chapter_no6.pdf

Gregory, W. J. (2018). Honeypots: Not for winnie the pooh but for winnie the pedo law enforcement's lawful use of technology to catch perpetrators and help victims of child exploitation on the dark web. *George Mason Law Review, 26*(1), 259-[ii].

Haaszf, A. (2016). Underneath it all: Policing international child pornography on the dark web. *Syracuse Journal of International Law & Commerce, 43*(2), 353–380.

Hadjimatheou, K. (2022). Policing the dark web: Ethical and legal issues. In *European Union's Horizon 2020. International conference on computer and communications (ICCC), 2020*, pp. 1694–1705. https://doi.org/10.1109/ICCC51575.2020.9345271

Jadoon, A., Iqbal, W., Amjad, M., Afzal, H., & Bangash, Y. (2019). Forensic analysis of tor browser: A case study for privacy and anonymity on the web. *Forensic Science International, 299*, 59–73. https://doi.org/10.1016/j.forsciint.2019.03.030

Jardine, E. (2015). *The dark web dilemma: Tor, anonymity and online policing*. Centre for International Governance Innovation and Chatham House.

Joh, E. E. (2009). Breaking the law to enforce it: Undercover police participation in crime. *Stanford Law Review, 62*, 155.

Kaur, S., & Randhawa, S. (2020). Dark web: A web of crimes. *Wireless Personal Communications, 112*(4), 2131–2158. https://doi.org/10.1007/s11277-020-07143-2

Kerr, J. (2018). How can legislators protect sport from the integrity threat posed by cryptocurrencies? *International Sports Law Journal, 18*, 79–97. https://doi.org/10.1007/s40318-018-0132-0

Kerr, O. S., & Murphy, S. D. (2017–2018). Government hacking to light the dark web: What risks to international relations and international law. *Stanford Law Review Online, 70*, 58–69.

Kumar, A., & Rosenbach, E. (2019). The truth about the dark web: Intended to protect dissidents, it has also cloaked illegal activity. *Finance & Development, 56*(3), A007. Retrieved 23 December 2021, from https://www.elibrary.imf.org/view/journals/022/0056/003/article-A007-en.xml

Leukfeldt, R., & Holt, T. (2020). *The human factor of cybercrime* (1st ed.). Routledge. Retrieved 17 December 2021, from https://www.taylorfrancis.com/books/edit/10.4324/9780429.460593/human-factor-cybercrime-rutger-leukfeldt-thomas-holt

Priyadarshan, J. (2022). Dark web: Criminal activities and the way forward. *The Law Chapterer*. Retrieved 4 January 2022, from https://thelawchapterer.in/2020/06/07/dark-web-criminal-activities-and-the-way-forward/#_ftnref9

Romeo, A. (2016). Hidden threat: The dark web surrounding cyber security. *Northern Kentucky Law Review, 43*(1), 73–86.

Rong, C., & Yang, G. (2003). Honeypots in blackhat mode and its implications [computer security]. In *Proceedings of the fourth international conference on parallel and distributed computing, applications and technologies* (pp. 185–188). IEEE. https://doi.org/10.1109/PDCAT.2003.1236284

Schein, D. D., & Trautman, L. J. (2020). The dark web and employer liability. *Colorado Technology Law Journal, 18*, 49.

Vendius, T. T. (2015). Proactive undercover policing and sexual crimes against children on the internet. *European Review of Organised Crime, 2*(2), 6–24.

Webber, C. (1995). Whistleblowing and the whistleblowers protection bill 1994. *Auckland University Law Review, 7*(4), 933–958.

Wells, M., Finkelhor, D., Wolak, J., & Mitchell, K. J. (2007). Defining child pornography: Law, enforcement dilemmas in investigations of internet child pornography possession. *Police Practice and Research, 8*(3), 269–282.

Zhang, H., & Zou, F. (2020). A survey of the dark web and dark market research. In *2020 IEEE 6th international conference on computer and communications (ICCC),* (pp. 1694–1705). https://doi.org/10.1109/ICCC51575.2020.9345271

Insights into the Next Generation of Policing: Understanding the Impact of Technology on the Police Force in the Digital Age

Shasha Yu and Fiona Carroll

Abstract Technological advancements have caused major changes in policing, in fact, we can say that it has transformed police work in the twenty-first century. Technology has made policing more efficient, productive and enabled them with new ways to keep safe whilst also communicate and interact. However, whilst most of these technological advancements have been beneficial to the police, some features continue to create adverse effects (i.e. these have created many new types of digital/cybercrimes) which are creating immense challenges for the police force. One of these challenges is the increased need to understand and use digital evidence. Police are now required to keep up with the fast-evolving technological world in order to catch criminals, protect society and reduce crime. This chapter gives a glimpse into the changes in the police environment with a particular focus on the application of artificial intelligence (AI) and big data in policing. It will share insights around the challenges of digital policing such as public security and individual rights; transparency of evidence and the black box effect; efficiency and legality of law enforcement; and emerging crimes and lagging laws. Finally, this chapter discusses the digital police future and all the potential benefits of predictive policing, digital collaboration and VR training, to name a few.

Keywords Digital policing · Artificial intelligence · Big data · Predictive policing · Digital crimes

S. Yu (✉)
Cardiff School of Technologies, Cardiff Metropolitan University Llandaff Campus, Cardiff, UK

School of Professional Studies, Clark University, Worcester, MA, USA
e-mail: ShaYu@clarku.edu

F. Carroll
Cardiff School of Technologies, Cardiff Metropolitan University Llandaff Campus, Cardiff, UK
e-mail: fcarroll@Cardiffmet.ac.uk;
https://www.cardiffmet.ac.uk/technologies/staff-profiles/Pages/Fiona-Carroll.aspx

R. Montasari (ed.), *Artificial Intelligence and National Security*,
https://doi.org/10.1007/978-3-031-06709-9_9

1 Introduction

In 1829, Sir Robert Peel, known as the 'father of modern policing', founded the Metropolitan Police Service in London, which is considered to be the beginning of modern policing (Jenkins, 1998). For two centuries, Peel's principles have had a profound impact on the development of policing in all countries; community-based policing has brought a sense of security to people all over the world. However, the last two decades have seen sweeping social, economic and technological changes. Technological innovations represented by artificial intelligence, 5G networks and big data have had a huge impact on the way people work and live. They have placed new and higher demands on policing in this new era. Indeed, new forms of crime are constantly emerging, challenging the ability of police teams to respond. The popularity of electronic devices such as mobile phones, tablets and computers has led to expectations for police teams to provide more online police services. Moreover, crime networks are rapidly expanding in a borderless manner, making cross-regional and international police cooperation a priority. The gradual increase in AI driven crime, blockchain crime, cybercrime and other digital crimes has led to the need to improve the overall technological literacy of police teams. Recruiting and training more technical experts has become an urgent concern for the police forces.

Specifically, the development of network and encryption technology has made crimes and criminals' identities more hidden and more costly and technically difficult to solve (Anderson et al., 2013). Also, the popularity of the Internet of Things and the emergence of blockchain has made traditional police investigation and evidence collection methods face huge challenges (Kearns & Muir, 2019). In addition, the availability of hacking software has made the cost and threshold of crime increasingly low, but the cost of solving crimes increasingly high (Markoff, 2016). At the same time, the police force also faces many constraints: the legal system is relatively backward compared to technology, making the basis for law enforcement insufficient which ties the hands of the police. The budgetary constraints limit the overall operational capability of the police and the development of technical equipment (Kearns & Muir, 2019). Also, the conflict between law enforcement powers and civil rights creates moral dilemmas (with administrative and judicial implications) for police work using advanced technology. This chapter will reflect on these new technological advancements, and in particular, how the police force is responding to the challenges of this evolving digital age.

2 Changes in the Policing Environment in the Era of Big Data and Artificial Intelligence

2.1 *Digitisation of Work Models*

The community policing model currently used in many countries originated in the 1960s and focuses on developing relationships with community members, concentrating on problem solving through active public participation, partnership and prevention (Koslicki et al., 2021; Rutland, 2021). It is the authors' opinion that with rapid technological advances, the traditional community policing model alone is far from adequate for modern policing. The development of the internet has changed the way people interact from mainly 'face-to-face' to 'end-to-end'. Analysis shows that internet users have more than doubled in the last decade, climbing from 2.18 billion at the beginning of 2012 to 4.95 billion at the beginning of 2022, with the 'typical' global internet user now spending nearly 7 hours a day using the internet on all devices (Kemp, 2022). This popularity of the internet has led to expectation of policing as a public service moving from offline to online (Kearns & Muir, 2019). The police are also now having to deal with a public that is accustomed to living online. This is driving the police to provide more channels for digital services. The introduction of various digital policing tools has freed police officers from monotonous and repetitive tasks, and even taken on some of the specialist tasks, increasing the efficiency of the force. Furthermore, the digitisation of policing processes and the resulting explosion of policing data have made it possible for predictive policing based on data analysis. This enables policing to move from reactive to proactive.

2.2 *Expanded Field of Work*

Crime in the traditional sense has always been associated with a specific location, victim and offender; however, the development of the internet has broken this concept. In an increasingly digital society, there are more and more cases where the perpetrators and victims are trans-regional and even transnational. Crimes are more likely to involve networks that span multiple jurisdictions, and relevant digital data is sometimes only available in other jurisdictions, thus requiring national and international cooperation to access such data.

In addition, the metaverse, marked by Meta, has opened the door for the virtual world to connect with the real world (Facebook, 2021). Digital identities and digital assets in the virtual world are increasingly becoming objects and targets for criminals to work with. This means that the field of work in the police is expanding from the real world to the virtual world. Furthermore, the rapid development of the Internet of Things and blockchain has led to profound changes in the way police investigate and collect evidence, with more evidence only available digitally. Police increasingly need access to data held by various connected devices, increasing the need to collaborate with third parties. Blockchain technology and cryptocurrencies take banks and other financial institutions out of the picture with decentralised, anonymous internet data recording, making digital forensics more challenging for the police.

2.3 Increased Technical Requirements for Police Officers

In the age of big data, data is the new oil. High-quality police data from various departments is the basis for the healthy development of digital policing. This puts forward higher technical requirements for all police officers, who need to be skilled in the use of various police systems. Also, they need to fully understand the principles of system operation and the meaning of data, in order to avoid 'rubbish in, rubbish out' in the collection of data.

In today's increasingly technologically advanced world, digital crime is simultaneously moving towards low barriers and high technology (Horsman, 2017). On the one hand, the low cost of software replication has made some criminal software readily available and users do not even need to have advanced cyber skills to engage in sophisticated forms of cyber-attack and fraud. This leads to a rapid growth in the number of digital crimes. For example, Blackshades, a software described as a 'criminal franchise in a box', was sold for as low as $40 through PayPal in 2014 (Markoff, 2016). This became prevalent, allowing users with no technical skills to effectively deploy ransomware and perform eavesdropping operations (Markoff, 2016). This was only later discontinued after a major international cyber crackdown (Sullivan, 2015). Furthermore, due to the highly open nature of AI systems, many new AI algorithms can be independently replicated by other researchers within months (Markoff, 2016). This is the case even when the source code has not been publicly shared, thus reducing costs and making even the latest technology easily available for criminal purposes (Hayward & Maas, 2021). On the other hand, technological updates have put enormous pressure on the development and application of anti-crime systems. The police need to work with top technological research institutes, private companies, etc. to keep up with the requirements of technological development. Also, they need to develop a pool of technical personnel through police recruitment and on-the-job training to increase the technical strength of the force.

3 Current Status of the Application of Artificial Intelligence and Big Data in Policing

3.1 Combating Crime

The fight against crime has traditionally been one of the most important responsibilities of the police and involves many highly specialised tasks. Fortunately, thanks to technological developments, some of these tasks can now be carried out more effectively with intelligent systems.

Based on computer image recognition and deep learning technology, it is possible to collate and classify crime scene photographs and find clues from them that can help solve crimes (Abraham et al., 2021; García-Olalla et al., 2018). Also, it is now possible to match objects from crime scenes with objects from previous incidents and search for links to other crimes (Saikia et al., 2017). Moreover, the technology can also be used to discover images of child sexual abuse and find missing persons by analysing photographs, CCTV footage, evidence documents and crime records. The Knowledge Discovery System (Ozgul et al., 2011) creates knowledge from structured and unstructured data sources that can be easily represented in an inferential manner. It has been used by nearly two hundred law enforcement agencies across the USA to search the internet for clues, pointing to victims of human trafficking and sex trafficking to track down potential victims (Baraniuk, 2019). It is being tried in investigations of drugs, illegal weapons sales and counterfeit goods (Baraniuk, 2019). Machine learning can also help forensic scientists to reconstruct faces from unidentified skulls. The system reassembles pieces of skulls from 3D scans and creates thousands of 3D reconstructions to fill in the missing pieces, it then searches to find a match to generate a 3D face to help identify the victim (Baraniuk, 2019; Liu & Li, 2018).

Crime prediction is one of the most important application of AI in policing. Several empirical studies on high-risk crime areas and crime trends have been conducted in the USA (e.g. Chicago (Yuki et al., 2019); YD County (Wu et al., 2018); Columbus, Ohio; and St. Louis, Missouri (Rock et al., 2008)) and Brazil (e.g. Rio de Janeiro (Aguirre et al., 2019)) and China (e.g. ZG City (Yu et al., 2020)). Some of these show a high accuracy (over 90%) in predicting crime, or improving police efficiency. In addition, studies that predict crime trends and crime types (across years) based on an individual's history of criminal charges (Chun et al., 2019) have also contributed to crime prevention. Indeed, predictive crime systems have been shown to be effective in many places and for many scenarios. For example, the New York City Police Department's (NYPD) Domain Awareness System (DAS) (Levine et al., 2017), developed in 2008, informs decision-making through pattern recognition, machine learning and data visualisation. It provides analytics and customised information to police officers' smartphones and police desktops. The system has been effective in combating terrorism and improving the effectiveness in fighting crime, with the city's overall crime index decreasing by 6% since the department-wide deployment of DAS in 2013 (Levine et al., 2017). In addition, DAS saves an estimated $50 million annually by increasing the efficiency of NYPD staff (Levine et al., 2017).

Police assistance systems have the capability to leverage the high precision of machine sensing devices, the high performance of computers and the high analytical power of artificial intelligence systems. They do this to assist police experts with cognitive technology and machine learning. For example, AI-based facial emotion recognition, AI verbal emotion recognition and deception recognition software have shown great potential for police interrogation. In detail, they have the ability to capture precise details, sophisticated rational analysis and help to eliminate bias in the criminal justice and legal systems (Noriega, 2020). One study analysed changes in human emotions by using computer-based algorithms to analyse the colour of human facial blood flow with 90% accuracy, compared to 75% accuracy for a control group of humans (Benitez-Quiroz et al., 2018). Another study found that as a person's mood changes, the tonal parameters change accordingly (Dasgupta, 2017). By extracting features such as micro-expressions and audio and analysing responses in real time, AI-assisted interrogation systems have shown good performance (Noriega, 2020). Crime scene image classifiers, on the other hand, liberate forensic examiners from the repetitive, laborious and time-consuming processing of crime scene images through computer vision and machine learning (Abraham et al., 2021).

3.2 *Serving Society*

According to the WHO (2018), 1.35 million people die and more than 50 million are seriously injured in road traffic accidents or other traffic-related incidents worldwide each year. Most road traffic accidents are not accidental, but predictable and controllable, and AI-based traffic management systems can be effective in reducing traffic accidents. For example, in 2017, an AI-based traffic management pilot programme was launched in Nevada, USA (DOT.GOV, 2020). The system is based on deep learning technology that identifies and predicts traffic patterns related to collisions and congestion based on connected vehicle data, cameras, roadside sensors and other traffic-related data to help officers implement countermeasures. The one-year pilot programme showed a 17% reduction in major crashes and an average response time of 12 minutes earlier (DOT.GOV, 2020).

Furthermore, with the increase in car ownership, road traffic congestion has become a problem for people. With the help of technologies such as artificial intelligence, cloud computing, big data, mobile internet, GPS and GIS, intelligent traffic systems (ITS) are playing an increasingly important role in traffic forecasting and scheduling (Li et al., 2021). It enables intelligent traffic management and assists police officers in traffic analysis, incident sensing, command and dispatch, police disposal, integrated communications and emergency management (Zhengxing et al., 2020). ITS can provide the public with detailed information on road construction, congestion, accidents and traffic control measures. It can provide traffic guidance information to the public via variable message signs (VMS) and the internet, helping drivers to be able to plan their routes and facilitate people's travel (Li & Lasenby, 2022).

The artificial intelligence-based crime hotspot query and prediction system (Wang et al., 2020) can predict the impact of urban planning adjustments on crime rates based on historical data. Thus, enabling construction planning departments to adjust their plans in time to reduce urban crime rates. The system visualises the crime rate of a given map and generates a map of crime hotspots in the area as a basis for adjusting urban planning and building design (Wang et al., 2020). By inputting changes to the building design, it can detect whether the aim of reducing the urban crime rate has been achieved and revalidates the adjusted map.

The creation of a dangerous event detection and warning systems through artificial intelligence is also playing an active role in serving the community. A shooting warning system called ShotSpotter (Meijer & Wessels, 2019), which uses sensors to detect gunshots, analyses data and immediately relays it to police, can detect more than 90% of shootings with precise location in less than 60 seconds. This significantly improves response (caqb.gov, 2020). For some areas with a high occurrence of shootings, the system enables officers to accurately determine the location of shots fired and to arrive on scene more quickly (Meijer & Wessels, 2019). South Africa, one of the countries with the highest murder rates in the world, has increased the recovery rate of illegal guns fivefold after adopting the technology (Shaw, 2018).

Finally, facial recognition technology not only plays an important role in fighting crime and hunting down fugitives, but is also used to find missing persons. The Indian Police in the southern state of Telangana launched a new facial recognition app in 2020 to reunite thousands of missing and trafficked children with their families (Nagaraj, 2020). The app can identify up to 80 points on a person's face to find a match and can match one million records per second (Nagaraj, 2020).

4 Insights Around the Challenges of Digital Policing

4.1 Conflict Between Public Safety and Individual Rights

Data collection poses privacy risks Police support systems based on artificial intelligence technology usually require large amounts of data to support continuous improvement and increase the accuracy of analysis results. Much of this data comes from government departments' information systems and data collection devices such as surveillance cameras installed in public places. The proliferation of the Internet of Things and wearable devices has enabled many private organisations to capture large amounts of data, which also provides an important source of data for the police.

How this data is managed is a double-edged sword. On the one hand, the large amount of data effectively leverages the effectiveness of digital policing systems. It allows the police to act as a more accurate prognosticator and quicker responder based on big data analysis. It provides them with a powerful way to fight crime, serve the community and provide a sense of security to the public. On the other hand, the ubiquitous surveillance network built to capture data poses a risk to personal privacy exposure. Some biometric information such as voiceprint, gait, fingerprints and faces can be captured without people being aware of it, resulting in personal privacy breaches. Facial recognition systems can accurately identify people even if their faces are wearing masks (Singh et al., 2017). Moreover, high pixel cameras can even identify faces and license plates in photographs taken from several kilometres away (Schneier & Koebler, 2019). When people's behaviour is under surveillance, it may result in further privacy violations. Aware of this, people may reduce their behaviour in places under surveillance. The resulting chilling effect may also trigger self-censorship, thereby reducing the space for individual rights.

Artificial intelligence is highly capable of data aggregation and can infer surprisingly private information from seemingly irrelevant data. For example, scientists can use AI to distinguish homosexuals from heterosexuals from facial images alone (Wang & Kosinski, 2018). It can also explore the relationship between facial features and other elements, such as political views, psychological conditions or personality traits (Wang & Kosinski, 2018), which further exacerbates the risk of privacy breaches for people.

Machine bias leads to prejudice and discrimination Police prediction systems predict the likelihood of a case or policing risk occurring by analysing historical data. This enables police to deploy limited police resources more rationally to high-risk areas and domains, improving the efficient use of police resources. However, because of the sampling bias often presented at data collection and the resulting machine bias may cause the analysis of place specific areas, populations, and races to be put in a discriminatory position. In fact, by treating households in specific areas and specific human races as 'high risk', this can also erroneously lead to over-enforcement of some communities and neighbourhoods and under-enforcement of others. Over time, this affords to more data being collected on individuals and incidents in targeted communities than elsewhere, further reinforcing this bias and discrimination. A study based on the predictive policing tool PredPol (Obermeyer et al., 2019) showed that while drug offences were roughly the same across races, predictive policing targeted blacks at roughly twice the rate of whites, with non-white races receiving targeted policing at 1.5 times the rate of whites. The researchers emphasise that this finding applies broadly to any predictive policing algorithm that uses unadjusted police records to predict future crime (Obermeyer et al., 2019).

Inferential analysis can lead to distortions or misleading Artificial intelligence automatically explores data based on some kind of model and creates derived or inferred data to make predictions and responses accordingly. These derived or inferred data are collectively referred to as 'inferential analysis' and they are only possibilities rather than certainties (Wachter & Mittelstadt, 2019). Both Facebook and Google have experienced instances of AI labelling black videos or photos as 'primates' or 'chimpanzees' (Mac, 2021). Very often, the results of AI analysis may be only slightly wrong and therefore difficult to detect, but enough to distort the reasonable evaluation of people by others, with adverse effects or consequences (Ishii, 2019). For example, when someone is incorrectly identified as a fugitive, they are identified as high risk by the policing system and thus wrongly subjected to law enforcement distress.

4.2 Conflict Between Transparency of Evidence and Black Box Effect

Many predictive policing systems based on artificial intelligence can provide accurate intelligence on crime analysis and help the police fight crime. However, due to the black box effect of AI and the inability to provide convincing explanations and justifications for its results, the police can be faced with the embarrassing situation of insufficient evidence or non-compliance with legal norms.

Artificial intelligence automatically explores data and generates analytical results based on a certain model, but it is not clear to the user or even the programmer how these results are generated, which is known as the algorithmic black box (Rudin & Radin, 2019). For example, although Google DeepMind's *AlphaGo* beat its human opponent, it could not explain why it made the move, nor could humans fully understand or unravel its underlying principles (Haddadin & Knobbe, 2020). To address the problems posed by algorithmic black boxes, some companies and researchers have looked at employing explainable artificial intelligence (XAI) to generate explanations for AI recognition systems that explain the causes of the activity leading to a given classification (Das et al., 2021; Houzé et al., 2020). However, it has also been pointed out that many popular and highly cited XAI methods in the literature may not explain anything at all, as they may make complex explanations that users cannot understand or mislead them (e.g. by giving them unreasonable trust) (Kenny et al., 2021). Indeed, complex intelligence and simple explanations are inherently an oxymoron. There is a growing insistence that it is in principle impossible to explore the 'black box' of AI (Innerarity, 2021).

Article 22 of the General Data Protection Regulation (GDPR, 2018) therefore provides that individuals shall not be subject to decisions which are based solely on automated processing (e.g. algorithms) and which are legally binding or which significantly affect them. The reality of policing is that officers may not have enough time or information to conduct secondary screening in their work, and thus tend to adopt the results of AI analysis directly. In particular, the more accurate the system, the more likely it is to be trusted, to the extent that officers do not have the confidence to use their own judgement to challenge or refute the course of action suggested by the algorithm, thus abandoning professional intuition in favour of machine results (Kearns & Muir, 2019). This essentially goes against the legal requirement for decision makers to consider all relevant factors and information when making decisions. Given the potential for systematic bias in AI systems, law enforcement actions based on biased analysis results can in fact develop machine bias into biased law enforcement.

4.3 Conflict Between Efficiency and Legality of Law Enforcement

In contrast to the rapid growth in demand for various policing responses, there has been limited growth in the size of the police force and even cuts in police funding (Kearns & Muir, 2019). In this context, the use of advanced technology to support law enforcement is undoubtedly the best option. The proliferation of new technologies has made this possible and has been shown to significantly improve the efficiency of the police force. However, as administrative authorities, the police should be authorised and regulated by law to enforce the rights of the public, whereas the rapid pace of technological development has kept this authorisation lagging behind the need of law-enforcement.

Legality issues in the adoption of technology Live facial recognition (LFR) is a controversial identification technology. It uses face recognition technologies to identify faces captured on near real-time video images and then matches these faces to a police watch list of individuals, which is often used by police to identify fugitives, unknown persons and so on (Bradford et al., 2020). Opponents argue that facial recognition is completely different from other forms of surveillance. Because the face (unlike fingerprints) is difficult to hide and can be scanned and recorded from a distance without knowledge, it threatens human rights and according to Hartzog and Selinger (2019) should be banned altogether. Indeed, California has banned the use of LFR by law enforcement agencies in 2019 (CA.GOV, 2019). In August 2020, the Court of Appeal ruled that the use of facial recognition technology by South Wales Police was unlawful (Rees, 2020). Technology providers such as Amazon, IBM and Microsoft have also announced that they are calling for a moratorium on the sale of LFR technology to police forces around the world (Amazon, 2020; IBM, 2020; Microsoft, 2020). In June 2021, the UK Information Commissioner's Office published 'The use of live facial recognition technology in public places', which provides opinion on the use of LFR (Office, 2021). However, the use of this technology is still not regulated in many police departments around the world.

Automatic Number Plate Recognition (ANPR) is another controversial recognition technology that uses optical character recognition to read vehicle number plates (Patel et al., 2013). This can be used to check whether a vehicle is registered or licensed, as well as to create vehicle location data. The UK's ANPR database holds up to 20 billion records, with around 25 to 40 million plates being scanned every day, and this data is stored for up to 12 months (Kearns & Muir, 2019). But a citizen tracking system of this magnitude has never had any statutory authority since its creation (Kearns & Muir, 2019). In order to improve efficiency, police in various countries are gradually adopting some other AI-based policing prediction systems. However, scholars have pointed out that the challenge with AI algorithms designed to predict crime is that it may undermine the jurisprudential principle that a person can only be punished for crimes already committed, not for possible future crimes predicted by the AI (Broadhurst et al., 2019).

Legitimacy issues in data collection A widely adopted method to improve the efficiency of law enforcement is where the police and private entities/organisations share data, hardware, software or intelligence, which is referred to as a Public-Private Collaboration (P-PC) (GOV.UK, 2021). However, the impact that private organisations have on the legitimacy of police data in collaborations has also become a concern (GOV.UK, 2021). Some private manufacturers do not provide their algorithms and training datasets for commercial confidentiality protection purposes, leading to court challenges to their legality (GOV.UK, 2021). In addition, the extensive collaboration between the private sector and the police has raised questions about its impact on police enforcement. For example, *Axon Enterprise* (formerly *Taser International*) provides body cameras to 47 of the 69 largest police agencies in the USA. It has trained 200,000 officers to collect 30 petabytes of video (10 times larger than the Netflix database) using AI systems. This raises concerns about whether private surveillance providers are having an impact on police departments' investigations and arrests by exerting undue influence (Hayward & Maas, 2021).

Secondly, as policing becomes increasingly reliant on electronic evidence, it is becoming increasingly challenging to discern the authenticity of electronic evidence captured from external sources, and even the most experienced experts can be fooled (Wu & Zheng, 2020). In a study on the use of artificial intelligence to tamper with medical images, researchers successfully fooled three radiologists and the latest lung cancer detection AI by automatically injecting or removing lung cancer information from a patient's 3D CT scan using artificial intelligence (Mirsky et al., 2019). An important admissibility criterion for electronic evidence is that it has authenticity and integrity, and its examination runs through the entire process of electronic evidence from its generation to its final submission to the court (Wu & Zheng, 2020). And as a result of the easily tampered nature of electronic evidence and the lack of necessary technical knowledge, it may be difficult for the police to endorse the authenticity and integrity of electronic evidence from external data sources. This would make the evidence potentially defective in terms of legitimacy.

4.4 Conflict Between Emerging Crimes and Lagging Laws

AI-enabled crimes With the rapid development of various AI technologies, more and more AI crimes have emerged and are still emerging, but the study of AI criminality and related law is still an emerging field. Hayward and Maas (2021) classify AI crimes into three categories: crimes with AI, crimes on AI, and crimes by AI. Within each category there are emerging forms of crime that require attention.

1. Crimes with AI, refers to crimes in which AI is used as a tool (Hayward & Maas, 2021), such as committing fraud through AI-synthesized audio and video, and creating fake news to manipulate political elections. For example, in 2019, the CEO of a British energy company was asked by fraudsters to transfer €220,000 to a bank account in Hungary using audio Deepfake technology to imitate the voice of the CEO of its parent company (Damiani, 2019).
2. Crimes on AI, which refers to the use of AI as an attack surface (Hayward & Maas, 2021). This approach manipulates the outcome of an AI system or to evade AI detection by attacking it or tampering with the data to undermine its effectiveness in order to achieve the criminal's goal. Examples include manipulating a self-driving car to carry out a terrorist attack, or masquerading as a trusted user to bypass a biometric authentication system and carry out a hack.
3. Crimes by AI, which refers to the use of AI as an intermediary (Hayward & Maas, 2021). Some people with ulterior motives use autonomous robots as criminal shields to commit criminal activities, such as using military robots to carry out terrorist attacks, or using intelligent trading agents to discover and learn to execute profitable strategies that amount to market manipulation.

As AI plays a different role in AI-enabled crime, the legal liability that should be imposed on itself and its stakeholders become a focus of attention and difficulty. Hallevy (2010) proposes three legal models of criminal liability for AI entities. The first is the 'Perpetration-via-Another Liability Model', which treats the AI as an innocent agent and attributes liability to the entity that directed it to commit the crime. The second is the 'Natural-Probable-Consequence Liability Model', whereby users and developers are liable if they know that committing a criminal offence is a natural and probable consequence of the AI. The third is the 'Direct Liability Model', where the AI should be held directly liable as long as it satisfies the physical and mental elements of the crime.

The first model of liability described above applies in the vast majority of crimes where AI is used as a tool. However, due to the black box effect, it is difficult for any developer of AI to explain the results of its operation. Also, it is difficult to make a decision in a forensic investigation as to whether the crime was caused by a failure of the system itself or by malicious use by others. Such investigations will be even more difficult, especially when considering that large-scale AI systems are often the result of multi-party cooperation.

There are problems with forensic investigations where AI is used as the attack surface (Jeong, 2020). In the second model of liability, data forensics will be difficult because the AI system itself, as the target of the attack, has had its functionality and data compromised. Due to the ease of tampering with electronic data, it is difficult to ensure the authenticity and integrity of electronic evidence.

The third model of liability is established on the premise that AI has the legal status of independent liability. However, currently in the vast majority of countries AI has no formal legal status and in practice liability can only be transferred to the user of the AI or its creator. In this case, the more intelligent the product developed, the greater the risk borne by the developer, which would be detrimental to the advancement of technology. To address this issue, giving legal status to autonomous robots became the issue to be faced. In October 2017, a humanoid robot named Sophia was confirmed to have the citizenship of Saudi Arabia. In November 2017, Sophia was named the first non-human innovation champion named by the first United Nations Development Programme (UNDP) (Pagallo, 2018). This means that for the first time, humans have legally recognised that robots can have the same rights and responsibilities as natural humans.

Recognising the legal identity of AI will present serious challenges to the existing legal system. For example, when an AI commits a transnational crime (e.g. cyber fraud), countries will face many difficulties in terms of investigation, prosecution and extradition, as their legal status and liability are determined differently (Broadhurst et al., 2019). In April 2021, the EU published a proposal by the European Parliament and Council to establish uniform rules on AI (AI Act) and to amend certain EU legislation to regulate AI (EUROPA.EU, 2021), but there is still a long way to go from the proposal to the enactment and implementation of the law.

Virtual space crime The COVID pandemic, which began in 2020, has dramatically changed the way people work and live around the world. Over the past 2 years, we have witnessed more and more people moving their learning, shopping and socialising from offline to online. A number of major technology companies have also developed more smart home-based work-from-home technologies (Abril, 2021). Facebook launched a transformative metaverse development programme in 2021. This involves social, learning, collaboration and gaming and went live with Horizon Workrooms, the first generation of metaverse-based work platforms (Facebook, 2021). The global market for augmented reality (AR), virtual reality (VR) and mixed reality (MR) associated with the metaverse is forecasted to rise nearly tenfold from $30.7 billion in 2021 to almost $300 billion by 2024 (Alsop, 2021). This means far-reaching changes in the way people live and work in the near future, with more public and private activities taking place in virtual space. Along with this, crimes related to virtual spaces will increasingly become a challenge for the police to face.

Some pioneers are bringing real-world life into new virtual spaces. With technologies such as virtual reality and blockchain, artists can host private virtual concerts, parties and art gallery exhibitions in a metaverse. Investors can buy real estate and use it for rent or even get a mortgage (Shevlin, 2022). Blockchain technology company Tokens.com has even bought virtual land for a whopping $2.5 million to build a fashion neighbourhood and has attracted the likes of Louis Vuitton, Gucci, Burberry and other luxury brands (Kamin, 2021). Technologists believe that the metaverse will develop into a fully functional economy in just a few years, becoming as integrated into people's lives as email and social networks are today (Kamin, 2021).

The rapid growth of the metaverse means that the gap between virtual space and real space is rapidly closing. In some ways even creating hybrid spaces – such as buying virtual space items in real life, or buying real life items in virtual space – where almost everything that people enjoy in real life can be reproduced in virtual space. Virtual reality technology uses various hardware and software tools which interact with our senses (sight, hearing, touch, smell, sense of balance, etc.) to construct an environment that can be recognised as real by people's bodies and minds (Dremliuga et al., 2020). As a result, the legal issues involved in crimes committed in virtual space are as complex as in real life; there can be fraud, defamation, money laundering, theft of intellectual property, exchange of child abuse images, sexual harassment and even virtual rape (Strikwerda, 2015). In fact, the experience of being in a virtual environment can cause real harm to the body and mind. Studies have shown that people in VR may be afraid to walk across a plank of wood that is on the ground in the real world but is placed between roofs in virtual reality (Wang et al., 2019). VR customers may literally have a heart attack from a game (Cleworth et al., 2016). The forms of crime are even more complex due to the virtual nature of digital space and the ease of replication. Examples include infringement issues arising from the use of images of other people or deceased people as avatars.

The anonymous, remote and global nature of crimes in a virtual space will, no doubt, generate cross-border jurisdictional issues in crime investigations. Unlike real-life crimes which often have a clear place of occurrence, virtual space crimes are difficult to correspond to real-life jurisdictions through the place of occurrence in the virtual world (Dremliuga et al., 2020). And since most activities in the virtual world are carried out under pseudonyms and avatars, the real identity of the criminal is difficult to determine. It is also difficult to establish jurisdiction through the place where the criminal belongs. There are also difficulties with proof and extradition, as virtual space participants come from countries with different legal systems, and it is difficult for victims to apply for law enforcement in their own location (Dremliuga et al., 2020).

The second is the problem of evidence collection. Money in the digital world is cryptocurrency, as finance in the metaverse is powered by the blockchain (Kamin, 2021). A digital distributed public ledger that eliminates the need for a third party (such as a bank). This decentralised approach to transactions is highly secretive, making it difficult for the police to collect evidence of transactions.

5 The Future of Digital Policing

The three core ideas of policing established by Sir Robert Peel, the 'father of modern policing', are as relevant today as they were two centuries ago. As we move within a digital society, the goal of policing is more so than ever, to prevent crime. Indeed, effective police services reduce crime by shifting their work from reactive to proactive through predictive policing. The key to preventing crime is earning public support by strengthening policing collaboration with all parties and gaining external policing support. All parties need to share the responsibility for crime prevention, as if they had been given a volunteer police force. Finally, the police earn public support by respecting community principles. Indeed, gaining public support requires efforts to build reputation rather than force, and building a high quality police force is the foundation.

5.1 *Predictive Policing*

Predictive policing generally refers to 'police work that utilises strategies, algorithmic technologies, and big data to generate near-future predictions about the people and places deemed likely to be involved in or experience crime' (Sandhu & Fussey, 2021). Meijer and Wessels (2019) summarise the benefits and drawbacks of predictive policing based on an analysis of the literature on predictive policing from 2010 to 2017. The literature analysis shows that the potential benefits of predictive policing are that it helps to accurately deploy police forces to identify individuals who may be involved in criminal behaviour (Meijer & Wessels, 2019). In detail, only a limited number of studies focused on the effectiveness of predictive policing methods in practice. From Levine et al. (2017) in the New York Police Department (NYPD) and Mohler et al. (2015) in the Los Angeles and Kent Police Department, departments' empirical studies have shown positive results for predictive policing. But there are also some results indecisive, such as an experiment conducted by Hunt et al. (2014) in the Shreveport, Louisiana Police Department.

Meijer and Wessels (2019)'s analysis reveals that among the drawbacks claimed by academics for predictive policing are the lack of transparency in the models and the inability of law enforcement agencies to fully understand these algorithms (Datta et al., 2016; Schlehahn et al., 2015) to the extent that it is difficult to develop a fitting strategy (Townsley, 2017). Most forecasting models are data-driven rather than theory-driven, these models place too much emphasis on correlation rather than causation (Andrejevic, 2017). In addition, the forecasts are opaque and difficult to interpret (Chan & Bennett Moses, 2016), making the results difficult to apply. The lack of transparency in the prediction results may raise issues of accountability for police (Moses & Chan, 2018) and stigmatization of individuals (Schlehahn et al., 2015).

Moreover, Meijer and Wessels (2019) argue that the potential disadvantages mentioned above, although widely discussed, lack empirical evidence. The (limited number of) empirical evaluation studies have focused on testing whether the desired outcome is achieved, rather than whether this leads to adverse effects. The findings suggest that the actual effects of predictive policing still need to be further explored in policing practice. It is the authors' opinion that the techniques could be extended in the following ways:

1. Predictive policing is a policing technique that is still evolving and more empirical research is needed to enable its continuous improvement. More research

based on practical applications by frontline officers will help to continuously improve this technology.
2. Exploring predictive policing models based on a hybrid form of theory and data. Predictive policing systems based on machine learning excel at data analysis, but cannot make explanations. Predictive crime theories developed over a long period of policing practice, such as the routine activity theory (RAT), Pareto model, repeated victimization theory and near repeat theory and data models can complement this (Sieveneck & Sutter, 2021).
3. Enhanced training in the application of predictive policing systems. Predictive policing systems are only tools and their effectiveness depends on the extent to which the product is used correctly. Police departments should work with system developers to explore best practice and guide officers on how to use the system as a complement to their police duties, to reduce the practicalities of outright adoption or rejection by officers who do not understand the system (Alikhademi et al., 2021; Asaro, 2019). Police officers working with data collection should be properly trained on how to improve data quality to reduce bias and avoid 'rubbish in, rubbish out'. Police officers who apply predictive results should understand how the system works and its limitations in order to select appropriate application scenarios.
4. Combining predictive policing with crime prevention to reduce crime rates. Literature shows that the current use of predictive policing outcomes is mostly focused on interventions on a case-by-case basis and less on the prevention of crime trends as a whole. The combination of predictive policing outcomes with crime prevention techniques such as situational crime prevention (SCP),[1] may have a positive effect on crime reduction. It achieves this by assuming that offenders are rational beings and exploring the temporal and spatial aspects of crime causing events and the dynamics of the situation in an effort to create an unfavourable environment to reduce the chances of a criminal event occurring (Ho et al., 2022). SCP techniques have been widely adopted by many governments in the field of traditional crime, such as the UK, New Zealand, Canada, Scandinavia and the Australian Government. SCP has been successful in combating traditional crime, but further interdisciplinary research is needed in the area of cybercrime.

5.2 Collaboration in Policing

With crime moving in the direction of networking, digitalisation, intelligence and internationalisation, there is an urgent need for the police to strengthen cooperation with all parties. Also, there is a need to make full use of various resources such as

[1] This was originally developed in the 1970s by researchers in the UK Home Office research department (Clarke, 2009).

technology, laws and conventions to enrich the police force in order to cope with the rapidly growing demand for policing.

Many countries need legal equivalence before they can provide mutual legal assistance (MLA) to another country. Indeed, jurisdictions differ in their legal definitions of offences and the evidence that must be assessed to establish them. As a result, reaching consensus on the legal issues involved in transnational crime through international conventions is a necessary basis for international police cooperation. The United Nations Convention against Transnational Organized Crime (UNTOC), which came into being in 2000, established a mechanism for cooperation in dealing with transnational crime. UNTOC has become a basic guide for international police cooperation and has played an important role in combating transnational organised crime over the years. Moreover, the 2001 Council of Europe (CoE) Convention on Cybercrime (Budapest Convention), the world's first international convention on cybercrime, has been adopted by 67 countries around the world, including all 47 in Europe.

The authors feel that continued sharing and access to data on crime and criminals through intergovernmental organisations and increased international police exchange and cooperation are important ways to combat transnational crime. The International Criminal Police Organization INTERPOL (2022), the world's largest police organisation, has 195 member countries, each of which has an INTERPOL National Central Bureau (NCB). INTERPOL manages 19 police databases containing information on crimes and criminals (from names and fingerprints to stolen passports), linking the various national/regional databases in real time through a communication system called I-24/7 (INTERPOL, 2022). Peer to Peer enables police officers, even between countries without diplomatic relations, to work directly with their counterparts on investigations, evidence collection, analysis and the search for fugitives, particularly in the three most pressing areas of global crime: terrorism, cybercrime and organised crime (INTERPOL, 2022).

On the other hand, P-PC remain an important way for the police to continue to receive support. In 2012, the UK Home Office published 'Statutory Guidance for Police Collaboration', which sets out in detail the legal framework for policing, decision process for collaboration, legal requirements for collaboration, accountability, governance by policing bodies, powers of the secretary of state, models of collaboration, funding, workforce arrangements, legal duties and liability for breach, and procurement, etc. (GOV.UK, 2012). This lays the foundation for regulating P-PC and ensuring its healthy development. The police should strengthen their liaison with the legal profession to promote the development of legal norms related to policing and to clarify the basis for law enforcement, especially in relation to legal issues arising from emerging crimes. Clear and specific laws will both help police officers to enforce the law accurately and build public trust in the police. It is the authors' opinion that the police should also strengthen cooperation with universities, research institutes and high-tech companies to keep abreast of new technologies. This would enable them to apply the latest research findings to policing practices and provide empirical data and case studies for the development of policing technologies. The support of the community is also an important factor in building a safer society. The results of predictive policing systems should be used for crime prevention through cooperation with all parties. This would enable the promotion of environmental design that reduces crime, enhancing information and cooperation with charitable organisations to avoid poverty-induced crime, etc.

5.3 *Reflections on Enhanced Policing*

In the digital age, people have higher expectations of the police. Police departments should take advantage of technological developments to build a highly qualified police force. It is the authors' view that there is a need to strengthen practical police training based on technologies such as virtual reality. Police is a high-stress, high-risk practical job that requires constant training whilst working to keep up with changes in their surroundings. Especially front-line officers who may face violent crimes need to reduce their own risks through good vocational training and practical work. Technologies such as virtual reality have the power to simulate various policing scenarios as required. This enables officers to quickly improve their practical experience during immersion training. Research has shown that students are more focused and motivated to learn during immersive VR training (Jeelani et al., 2020). VR-based forensic practice courses in particular can develop the ability of police officers to work independently by simulating a variety of different crime scene scenarios, providing a broader and more cost-effective solution than traditional methods (Dhalmahapatra et al., 2020).

Furthermore, the authors feel that it is important to train or introduce technical personnel who are proficient in computer and network technology. A study conducted in Australia on the challenges faced by specialist police cybercrime units reveals the current state of the police when faced with cybercrime (Harkin et al., 2018). On the one hand, the workload expected to be taken on by cybercrime units has escalated and accelerated. On the other hand, the resourcing of cybercrime units has not matched the growing demand and the level of skills and training within the units is often inadequate to address the increasing complexity in investigating cybercrime. It was found that this can place a disproportionate strain on the few tech-savvy staff within specialist units (Harkin et al., 2018). The rapid growth of cybercrime has created an urgent need for more expert talent within police departments.

6 Conclusion

Policing is an important task in maintaining public order and combating crime. With the rapid development of technology, how the police adapt to the new digital societies is a concern that police, all over the world, will have to face. As we have seen, it is necessary for the police to adapt to the digital mode of work, the expanding field of their work and the ever-increasing technological requirements. As discussed, they need to take on board technologies such as artificial intelligence and big data to fight crime and serve society. Police today need to find ways to deal with the four pairs of contradictions between public security and individual rights; transparency of evidence and the black box effect; law enforcement efficiency and legality; and finally between emerging crimes and lagging laws. Looking to the future, Sir Robert Peel's core ideas about policing still apply to policing in the digital age. In detail, using artificial intelligence and big data to our advantage to implement predictive policing will help us to be more reactive and proactive, to prevent crime and reduce crime rates. Also, garnering support and resources of all kinds through greater international, intraindustry and public collaboration will enable policing to evolve in a more robust manner. Finally, above all else, it needs to be the appropriate and effective use of technology to build a stronger and higher quality police force which will form the fundamentals for the future of our policing.

References

Abraham, J., Ng, R., Morelato, M., Tahtouh, M., & Roux, C. (2021). Automatically classifying crime scene images using machine learning methodologies. *Forensic Science International: Digital Investigation, 39*, 301273.

Abril, D. (2021). *Big tech is pushing smart home devices as the latest work-from-home tools*. https://www.washingtonpost.com/technology/2021/11/22/smart-home-devices-security-remote-workers/

Aguirre, K., Badran, E., Muggah, R., & Geray, O. (2019). *Crime prediction for more agile policing in cities – Rio de Janeiro, Brazil*. https://igarape.org.br/wpcontent/uploads/2019/10/460154_Case-study-Crime-prediction-formore-agile-policing-in-cities.pdf

Alikhademi, K., Drobina, E., Prioleau, D., Richardson, B., Purves, D., & Gilbert, J. E. (2021). A review of predictive policing from the perspective of fairness. *Artificial Intelligence and Law*, 1–17.

Alsop, T. (2021). *Augmented reality (AR) and virtual reality (VR) market size worldwide from 2016 to 2024*. https://www.statista.com/statistics/591181/global-augmented-virtual-reality-marketsize/#statisticContainer

Amazon. (2020). *We are implementing a one-year moratorium on police use of recognition*. https://www.aboutamazon.com/news/policy-newsviews/we-are-implementing-a-one-year-moratorium-on-police-useof-rekognition

Anderson, R., Barton, C., Böhme, R., Clayton, R., Van Eeten, M. J., Levi, M., Moore, T., & Savage, S. (2013). Measuring the cost of cybercrime. In *The economics of information security and privacy* (pp. 265–300). Springer.

Andrejevic, M. (2017). Digital citizenship and surveillance— to pre-empt a thief. *International Journal of Communication, 11*, 18.

Asaro, P. M. (2019). Ai ethics in predictive policing: From models of threat to an ethics of care. *IEEE Technology and Society Magazine, 38*(2), 40–53.

Baraniuk, C. (2019). *The new weapon in the fight against crime*. https://www.bbc.com/future/article/20190228-how-ai-is-helpingto-fight-crime

Benitez-Quiroz, C. F., Srinivasan, R., & Martinez, A. M. (2018). Facial color is an efficient mechanism to visually transmit emotion. *Proceedings of the National Academy of Sciences, 115*(14), 3581–3586.

Bradford, B., Yesberg, J. A., Jackson, J., & Dawson, P. (2020). Live facial recognition: Trust and legitimacy as predictors of public support for police use of new technology. *The British Journal of Criminology, 60*(6), 1502–1522.

Broadhurst, R., Maxim, D., Brown, P., Trivedi, H., & Wang, J. (2019). *Artificial intelligence and crime*. Available at SSRN 3407779.

CA.GOV. (2019, October). *Ab-1215 law enforcement: facial recognition and other biometric surveillance (2019–2020)*. Available at: https://leginfo.legislature.ca.gov/faces/billTextClient.xhtml?bill_id=201920200AB1215

caqb.gov. (2020). *Mayor Keller and Interim Police Chief Harold Medina highlight ShotSpotter technology*. https://www.cabq.gov/police/news/mayor-kellerand-interim-police-chief-harold-medina-highlight-shotspottertechnology

Chan, J., & Bennett Moses, L. (2016). Is big data challenging criminology? *Theoretical Criminology, 20*(1), 21–39.

Chun, S. A., Avinash Paturu, V., Yuan, S., Pathak, R., Atluri, V. R., & Adam, N. (2019). Crime prediction model using deep neural networks. In *Proceedings of the 20th annual international conference on Digital Government Research* (pp. 512–514).

Clarke, R. V. (2009). Situational crime prevention: Theoretical background and current practice. In *Handbook on crime and deviance* (pp. 259–276). Springer.

Cleworth, T. W., Chua, R., Inglis, J. T., & Carpenter, M. G. (2016). Influence of virtual height exposure on postural reactions to support surface translations. *Gait & Posture, 47*, 96–102.

Damiani, J. (2019). *A voice deepfake was used to scam a CEO out of $243,000*. https://www.forbes.com/sites/jessedamiani/2019/09/03/a-voicedeepfake-was-used-to-scam-a-ceo-out-of-243000/?sh=192aef622241

Das, D., Nishimura, Y., Vivek, R. P., Takeda, N., Fish, S. T., Ploetz, T., & Chernova, S. (2021). Explainable activity recognition for smart home systems. *arXiv preprint arXiv:2105.09787*.

Dasgupta, P. B. (2017). Detection and analysis of human emotions through voice and speech pattern processing. *arXiv preprint arXiv:1710.10198*.

Datta, A., Sen, S., & Zick, Y. (2016). Algorithmic transparency via quantitative input influence: Theory and experiments with learning systems. In *2016 IEEE symposium on Security and Privacy (SP)* (pp. 598–617). IEEE.

Dhalmahapatra, K., Das, S., & Maiti, J. (2020). On accident causation models, safety training and virtual reality. *International Journal of Occupational Safety and Ergonomics*, 1–17.

DOT.GOV. (2020). Ai-*based traffic management pilot program implemented near Las Vegas contributed to a 17 percent reduction in primary crashes.*https://www.itskrs.its.dot.gov/node/209172

Dremliuga, R., Prisekina, N., & Yakovenko, A. (2020). New properties of crimes in virtual environments. *Advances in Science, Technology and Engineering Systems, 5*(6), 1727–1733.

EUROPA.EU. (2021). *Proposal for a regulation of the European Parliament and of the Council laying down harmonised rules on artificial intelligence (Artificial Intelligence Act) and amending certain union legislative acts*. https://eurlex.europa.eu/legal-content/EN/TXT/?uri=CELEX%3A52021PC0206

Facebook. (2021). *Introducing horizon workrooms: Remote collaboration reimagined*. https://about.fb.com/news/2021/08/introducing-horizonworkrooms-remote-collaboration-reimagined/

García-Olalla, O., Alegre, E., Fernández-Robles, L., Fidalgo, E., & Saikia, S. (2018). Textile retrieval based on image content from CDC and webcam cameras in indoor environments. *Sensors, 18*(5), 1329.

GDPR. (2018). *Automated individual decision-making, including profiling*. https://gdpr-info.eu/art-22-gdpr/

GOV.UK. (2012). *Statutory guidance for police collaboration*. https://assets.publishing.service.gov.uk/government/uploads/system/uploads/attachment_data/file/117559/police-collaboration.pdf

GOV.UK. (2021). *Briefing note on the ethical issues arising from public–private collaboration in the use of live facial recognition technology*. https://www.gov.uk/government/publications/public-private-useof-live-facial-recognition-technology-ethical-issues/briefingnote-on-the-ethical-issues-arising-from-public-private%2D%2Din-theuse-of-live-facial-recognition-technology-accessible

Haddadin, S., & Knobbe, D. (2020). Robotics and artificial intelligence—the present and future visions. In *Algorithms and Law* (pp. 20–23). Cambridge University Press.

Hallevy, G. (2010). The criminal liability of artificial intelligence entities-from science fiction to legal social control. *Akron Intellectual Property Journal, 4*, 171.

Harkin, D., Whelan, C., & Chang, L. (2018). The challenges facing specialist police cybercrime units: An empirical analysis. *Police Practice and Research, 19*(6), 519–536.

Hartzog, W., & Selinger, E. (2019). Why you can no longer get lost in the crowd. *The New York Times*, 17.

Hayward, K. J., & Maas, M. M. (2021). Artificial intelligence and crime: A primer for criminologists. *Crime, Media, Culture, 17*(2), 209–233.

Ho, M. H., Ko, R., & Mazerolle, L. (2022). Situational crime prevention (SCP) techniques to prevent and control cybercrimes: A focused systematic review. *Computers & Security*, 102611.

Horsman, G. (2017). Can we continue to effectively police digital crime? *Science & Justice, 57*(6), 448–454.

Houzé, E., Diaconescu, A., Dessalles, J. L., Mengay, D., & Schumann, M. (2020). A decentralized approach to explanatory artificial intelligence for autonomic systems. In *2020 IEEE international conference on Autonomic Computing and Self-Organizing Systems Companion (ACSOS-C)* (pp. 115–120). IEEE.

Hunt, P., Hollywood, J. S., & Saunders, J. M. (2014). *Evaluation of the Shreveport predictive policing experiment*. RAND Corporation.

IBM. (2020). *IBM CEO's letter to congress on racial justice reform*. https://www.ibm.com/blogs/policy/facial-recognition-sunsetracial-justice-reforms/

Innerarity, D. (2021). Making the black box society transparent. *AI & Society, 36*(3), 975–981.

INTERPOL. (2022). *What is Interpol?*https://www.interpol.int/Who-weare/What-is-INTERPOL

Ishii, K. (2019). Comparative legal study on privacy and personal data protection for robots equipped with artificial intelligence: Looking at functional and technological aspects. *AI & Society, 34*(3), 509–533.

Jeelani, I., Han, K., & Albert, A. (2020). Development of virtual reality and stereopanoramic environments for construction safety training. *Engineering, Construction and Architectural Management*.

Jenkins, T. A. (1998). *Sir Robert Peel*. Macmillan International Higher Education.

Jeong, D. (2020). Artificial intelligence security threat, crime, and forensics: Taxonomy and open issues. *IEEE Access, 8*, 184560–184574.

Kamin, D. (2021). *Investors snap up metaverse real estate in a virtual land boom*. https://www.nytimes.com/2021/11/30/business/metaverse-realestate.html

Kearns, I., & Muir, R. (2019). *Data-driven policing and public value*. https://www.police-foundation.org.uk/2017/wp-content/uploads/2010/10/data_driven_policing_final.pdf

Kemp, S. (2022). *Digital 2022: Global overview report*. https://datareportal.com/reports/digital-2022-global-overview-report#:~:text=Global%20internet%20users%3A%20Global%20internet,of%20the%20world%27s%20total%20population

Kenny, E. M., Ford, C., Quinn, M., & Keane, M. T. (2021). Explaining black-box classifiers using post-hoc explanations-by-example: The effect of explanations and error rates in Xai user studies. *Artificial Intelligence, 294*, 103459.

Koslicki, W. M., Lytle, D. J., Willits, D. W., Brooks, R., et al. (2021). 'Rhetoric without reality' or effective policing strategy? An analysis of the relationship between community policing and police fatal force. *Journal of Criminal Justice, 72*(C), 101730.

Levine, E. S., Tisch, J., Tasso, A., & Joy, M. (2017). The New York city police department's domain awareness system. *Interfaces, 47*(1), 70–84.

Li, D., & Lasenby, J. (2022). Mitigating urban motorway congestion and emissions via active traffic management. *Research in Transportation Business & Management*, 100789.

Li, Y., Chen, Y., Yuan, S., Liu, J., Zhao, X., Yang, Y., & Liu, Y. (2021). Vehicle detection from road image sequences for intelligent traffic scheduling. *Computers & Electrical Engineering, 95*, 107406.

Liu, C., & Li, X. (2018). Superimposition-guided facial reconstruction from skull. *arXiv preprint arXiv:1810.00107*.

Mac, R. (2021). *Facebook apologizes after A.I. puts 'primates' label on video of black men*. https://www.nytimes.com/2021/09/03/technology/facebookai-race-primates.html

Markoff, J. (2016). As artificial intelligence evolves, so does its criminal potential. *New York Times*.

Meijer, A., & Wessels, M. (2019). Predictive policing: Review of benefits and drawbacks. *International Journal of Public Administration, 42*(12), 1031–1039.

Microsoft. (2020). *Microsoft bans facial recognition sales to police*. https://www.silicon.co.uk/e-innovation/artificial-intelligence/microsoft-bans-facial-recognition-police-345703

Mirsky, Y., Mahler, T., Shelef, I., & Elovici, Y. (2019). {CT-GAN}: Malicious tampering of 3d medical imagery using deep learning. In *28th USENIX Security Symposium (USENIX Security 19)* (pp. 461–478).

Mohler, G. O., Short, M. B., Malinowski, S., Johnson, M., Tita, G. E., Bertozzi, A. L., & Brantingham, P. J. (2015). Randomized controlled field trials of predictive policing. *Journal of the American Statistical Association, 110*(512), 1399–1411.

Moses, L. B., & Chan, J. (2018). Algorithmic prediction in policing: Assumptions, evaluation, and accountability. *Policing and Society, 28*, 806–809.

Nagaraj, A. (2020). *Indian police use facial recognition app to reunite families with lost children*. https://www.reuters.com/article/us-india-crimechildren-idUSKBN2081CU

Noriega, M. (2020). The application of artificial intelligence in police interrogations: An analysis addressing the proposed effect AI has on racial and gender bias, cooperation, and false confessions. *Futures, 117*, 102510.

Obermeyer, Z., Powers, B., Vogeli, C., & Mullainathan, S. (2019). Dissecting racial bias in an algorithm used to manage the health of populations. *Science, 366*(6464), 447–453.

Office, I. C. (2021). *The use of live facial recognition technology in public places*. https://ico.org.uk/media/2619985/ico-opinion-the-useof-lfr-in-public-places-20210618.pdf

Ozgul, F., Atzenbeck, C., Celik, A., & Erdem, Z. (2011). Incorporating data sources and methodologies for crime data mining. In *Proceedings of 2011 IEEE international conference on Intelligence and Security Informatics* (pp. 176–180). IEEE.

Pagallo, U. (2018). Vital, Sophia, and Co.—the quest for the legal personhood of robots. *Information, 9*(9), 230.

Patel, C., Shah, D., & Patel, A. (2013). Automatic number plate recognition system (ANPR): A survey. *International Journal of Computer Applications, 69*(9), 21–33.

Rees, J. (2020). *Facial recognition use by South Wales police ruled unlawful*. https://www.bbc.com/news/uk-wales-53734716

Rock, D. J., Judd, K., & Hallmayer, J. F. (2008). The seasonal relationship between assault and homicide in England and Wales. *Injury, 39*(9), 1047–1053.

Rudin, C., & Radin, J. (2019). *Why are we using black box models in AI when we don't need to? A lesson from an explainable AI competition*.

Rutland, T. (2021). From compromise to counter-insurgency: Variations in the racial politics of community policing in Montreal. *Geoforum, 118*, 180–189.

Saikia, S., Fidalgo, E., Alegre, E., & Fernández-Robles, L. (2017). Object detection for crime scene evidence analysis using deep learning. In *International conference on Image Analysis and Processing* (pp. 14–24). Springer.

Sandhu, A., & Fussey, P. (2021). The 'uberization of policing'? How police negotiate and operationalise predictive policing technology. *Policing and Society, 31*, 66–81.

Schlehahn, E., Aichroth, P., Mann, S., Schreiner, R., Lang, U., Shepherd, I. D., & Wong, B. W. (2015). Benefits and pitfalls of predictive policing. In *2015 European Intelligence and Security Informatics Conference* (pp. 145–148). IEEE.

Schneier, B., & Koebler, J. (2019). AI has made video surveillance automated and terrifying. *Vice*.
Shaw, N. (2018). *South Africa adopts new audio tech to find location of gunshots and immediately alert police*. https://globalnews.ca/news/4446595/southafrica-shotspotter-gun-tech/
Shevlin, R. (2022). *Digital land grab: Metaverse real estate prices rose 700*. https://www.forbes.com/sites/ronshevlin/2022/02/04/digital-land-grabmetaverse-real-estate-prices-rose-700-in-2021/?sh=704677277cdc
Sieveneck, S., & Sutter, C. (2021). Predictive policing in the context of road traffic safety: A systematic review and theoretical considerations. *Transportation Research Interdisciplinary Perspectives, 11*, 100429.
Singh, A., Patil, D., Reddy, M., & Omkar, S. (2017). Disguised face identification (DFI) with facial keypoints using spatial fusion convolutional network. In *Proceedings of the IEEE international conference on Computer Vision Workshops* (pp. 1648–1655).
Strikwerda, L. (2015). Present and future instances of virtual rape in light of three categories of legal philosophical theories on rape. *Philosophy & Technology, 28*(4), 491–510.
Sullivan, G. (2015). Scary things about the 'Blackshades' rat. *Washington Post*, 20.
Townsley, M. (2017). Crime mapping and spatial analysis. In *Crime prevention in the 21st century* (pp. 101–112). Springer.
Wachter, S., & Mittelstadt, B. (2019). A right to reasonable inferences: Re-thinking data protection law in the age of big data and AI. *Colum. Bus. L. Rev*, 494.
Wang, Y., & Kosinski, M. (2018). Deep neural networks are more accurate than humans at detecting sexual orientation from facial images. *Journal of Personality and Social Psychology, 114*(2), 246.
Wang, H., Wang, Q., & Hu, F. (2019). Are you afraid of heights and suitable for working at height? *Biomedical Signal Processing and Control, 52*, 23–31.
Wang, J., Hu, J., Shen, S., Zhuang, J., & Ni, S. (2020). Crime risk analysis through big data algorithm with urban metrics. *Physica A: Statistical Mechanics and its Applications, 545*, 123627.
WHO. (2018). *Global status report on road safety*. https://www.who.int/publications/i/item/9789241565684
Wu, H., & Zheng, G. (2020). Electronic evidence in the blockchain era: New rules on authenticity and integrity. *Computer Law & Security Review, 36*, 105401.
Wu, S., Wang, C., Cao, H., & Jia, X. (2018). Crime prediction using data mining and machine learning. In *International conference on Computer Engineering and Networks* (pp. 360–375). Springer.
Yu, H., Liu, L., Yang, B., & Lan, M. (2020). Crime prediction with historical crime and movement data of potential offenders using a Spatio-Temporal Cokriging Method. *ISPRS International Journal of Geo-Information*, 9(12). https://www.mdpi.com/2220-9964/9/12/732
Yuki, J. Q., Sakib, M. M. Q., Zamal, Z., Habibullah, K. M., & Das, A. K. (2019). Predicting crime using time and location data. In *Proceedings of the 2019 7th International conference on Computer and Communications Management* (pp. 124–128).
Zhengxing, X., Qing, J., Zhe, N., Rujing, W., Zhengyong, Z., He, H., Bingyu, S., Liusan, W., & Yuanyuan, W. (2020). Research on intelligent traffic light control system based on dynamic Bayesian reasoning. *Computers & Electrical Engineering, 84*, 106635.

The Dark Web and Digital Policing

Megan Wilmot McIntyre and Reza Montasari

Abstract The Dark Web is a largely unknown part of the Internet, harbouring a space for both evil and good actions as a result of its secretive and anonymous nature. Unlike the Surface Web, it is much harder to digitally police the Dark Web because of how censored and difficult to access personal information is in the deepest part of the Internet. It is the challenges posed by the Dark Web to digital policing that is of interest in this chapter. Based on the issues explored, specific recommendations have been made in an effort to minimise the detrimental effect the Dark Web has on policing operations. After an exhaustive review of the available literature, the key findings illustrate that a balance is needed between policing and protecting the Dark Web. The predominant recommendation is the need for international cooperation and the collaboration of law enforcement agencies across jurisdictions all working towards the same goal. The findings also allude to the idea that whilst digital policing needs an element of consistency to be effective, the methods utilised also need to be unpredictable by the criminals that work on the Dark Web.

Keywords Dark Web · TOR browser · Cybercrime · Digital policing · Law enforcement · Anonymity

1 Introduction

Since its creation, the World Wide Web and the resources that it provides have not stopped expanding. With new information constantly being added, the Internet has knowledge and answers to offer for every scenario or question possible (Davis & Arrigo, 2021). However, this only refers to what is accessible through a regular

M. Wilmot McIntyre (✉) · R. Montasari
School of Social Sciences, Department of Criminology, Sociology and Social Policy, Swansea University, Swansea, UK
e-mail: Reza.Montasari@Swansea.ac.uk
URL: http://www.swansea.ac.uk

R. Montasari (ed.), *Artificial Intelligence and National Security*,
https://doi.org/10.1007/978-3-031-06709-9_10

online search and does not account for the masses of content that has been purposefully concealed from users who do not have access to a subsection of the Internet, known as the Dark Web (Finklea, 2017). The secretive and anonymous nature of the Dark Web enables both legitimate and illegitimate functions to be exercised. Law enforcement agencies (LEAs) are particularly concerned with the illegal actions that are performed as a result of manipulating the Dark Web (Ibid.). Davis and Arrigo (2021) highlight that despite ongoing research, LEAs and other related bodies still do not have a solidified and complete understanding of the capabilities of Dark Web user. As a result, this severely compromises their ability to develop and implement legislation and ethical practice that will be able to contend with the ever-progressing nature of the Dark Web.

This chapter focuses on the dilemmas that arise for digital policing through the presences of the Dark Web. The subsequent objectives are to give a detailed explanation of some of the key challenges faced, by including examples of cases that have evidenced such issues. The chapter also offers practical resolutions that could be implemented to counteract the challenges that threaten digital policing. This has been achieved by widespread research and the examination of the existing literature that has facilitated a nuanced and well-rounded understanding of this complex dimension. The remainder of this chapter is structured as follows. Section 2 provides an overview, the necessary background information serving to offer context and, in turn, enabling a better understanding. This will include an explanation of what the Dark Web is, how it can be utilised, a brief history and how it can be accessed. The notion of digital policing will also be considered in this section, looking at definitions and the way in which it is undertaken. Section 3 offers a critical evaluation of the key elements, by means of which the Dark Web can hinder the act of digital policing. Section 4 provides a set of specific recommendations to help to address or mitigate the issues analysed in Sect. 3. Section 5 discusses and contextualizes the main findings, describes their significance, evaluates the research carried out, and discusses an avenue for further research. Finally, the chapter is concluded in Sect. 6, which summarises the findings and contributions of this study to the existing body of knowledge.

2 Background

To begin with it can be useful to consider what cyberspace looks like as a whole and then where the Dark Web resides within this. Whilst there is no single definition of cyberspace (Zhang et al., 2015), different academics have attempted to produce one. For example, Ning et al. (2018) talk of cyberspace as something that is independent of time, metaphorical, digital and abstract and is rooted within a network of globally linked computers and other related infrastructures. Similarly, the Cabinet Office (2011) refers to cyberspace as 'an interactive domain made up of digital networks that is used to store, modify and communicate information', which includes the Internet among other information systems that support the framework, businesses

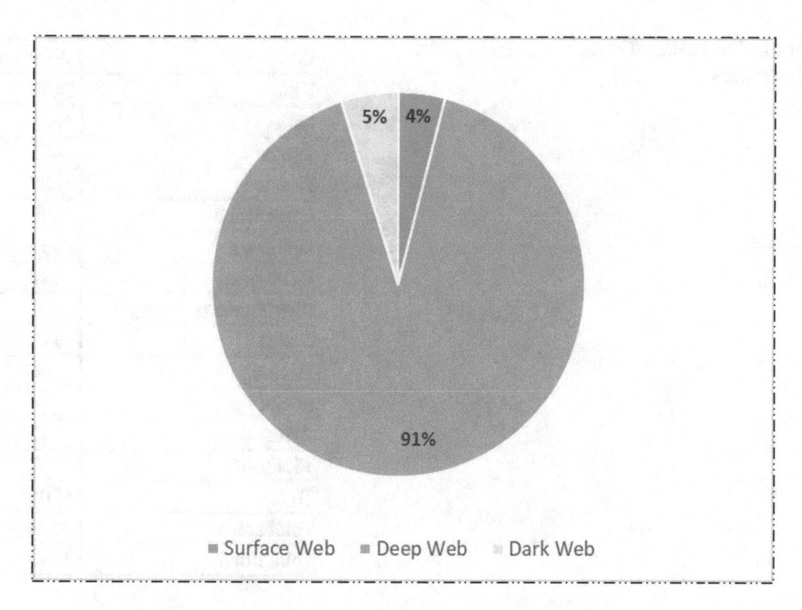

Fig. 1 Percentage size of subsections of the Internet

and services of the people. Cyberspace can be split into three separate parts using the analogy of an iceberg to illustrate this. The tip of the iceberg can be thought of as the Surface Web, just below this is the Deep Web and finally, further down at the bottom of the iceberg is the Dark Web (Cyware, 2019) as displayed in Fig. 1, which quantifies each subsection of the Internet into a percentage.

The Surface Web is made up of any information that is retrievable through regular search engines, like Google, meaning anyone with access to the Internet is able to view this content (Bright Planet, 2014). The Deep Web contains content that cannot be viewed through a standard search engine (Chertoff & Simon, 2015), as it is stored in a database that only generates results as a response to specific requests (Bergman, 2000); it can be accessed by using its URL (Uniform Resource Locators) and often requires a password as a form of protection (Cyware, 2019). The Dark Web, making up a small part of the Deep Web, contains content that has been made unreachable through normal search engines (Bright Planet, 2014).

Hence, a specific type of browser is necessary to access the Dark Web. There are various types; however, the most commonly referred to is The Onion Router (TOR) (Kaur & Randhawa, 2020). The TOR browser also functions in a way to protect users' identity when accessing content on the Dark Web; it does this by using 'relays' to transmit messages through, so the messages are not sent directly between two computers when visiting certain sites, but sent through multiple computers (Owen & Savage, 2015). By breaking up the direct connection, it makes it extremely difficult to link together all the connections and identify the user, offering anonymity for those using the TOR browser (Ibid.). The anonymous nature of the Dark Web has been taken advantage of for the purpose of carrying out a wide range of illegal

Table 1 Categories of Dark Web websites

Category	Websites
None	2482
Other	1021
Drugs	423
Finance	327
Other illicit	198
Unknown	155
Extremism	140
Illegitimate pornography	122
Nexus	118
Hacking	96
Social	64
Arms	42
Violence	17
Total	5205
Total active	2723
Total illicit	1547

activities; cybercriminals use the Dark Web to set up markets to buy and sell goods such as drugs, firearms, indecent images of children and stolen financial and personal details, to name a few (Kavallieros et al., 2021). Various types of hackers also use the Dark Web to gain profit and steal data, along with cyberterrorists who make use of the encrypted communication to share information globally and plan activities whilst remaining undetected (Ibid.).

Table 1 shows the results from research that aimed to scan and categorise websites found on the TOR network and conclude from the successfully classified websites that the most common uses for websites on the Dark Web are for criminal use (Moore & Rid, 2016). Whilst the Dark Web is largely used for illicit activities, there are many legitimate and legal uses that include ways for files or images to be stored more securely, either for personal use or in the context of whistle-blowers and journalists (Sui et al., 2015). This privacy also offers opportunities for authoritarian regimes of dictatorships to be exposed, thus protecting human rights and enabling freedom of expression (Chertoff, 2017), demonstrating the diverse ways in which the Dark Web can be used for both good and bad.

The early development of the Dark Web began around the same time as the initial creation of the Internet. In 1969 the Pentagon's Defence Advanced Research Project Agency (DARPA) developed the Advanced Research Project Agency Network (ARPANET) allowing messages to be sent between computers that were geographically separated. Within a few years, numerous secretive networks were set up in company with ARPANET with some going on to be branded as 'darknets' (McCormick, 2013). Fast forward to 2002, when researchers at the US Naval Research Labs (NRL), funded by DARPA, developed an early model of the TOR so that law enforcement officials could hide their identity on the Internet (Jacoby, 2016). However, the flaw here was that if only TOR users were law enforcement, it was easy to deduce that all TOR connections were for law enforcement purposes, making it unfit for use (Ibid.).

Thus, TOR needed to be made available to everyone so that TOR users could undeniably be anonymous; TOR was consequently released to the public in 2003 (Kushner, 2015). As a result of the continually advancing sphere of the Internet and the way in which it can facilitate crimes via the Dark Web, law enforcement has had to utilise certain methods to keep up and ensure that it is being effectively policed (Davis, 2017). Davis (2017) notes some of the main ways in which digital policing is carried out on the Dark Web, including the use of 'honeypot traps' which are traps set up by the police imitating websites harbouring illegal content that aim to obtain the IP (Internet Protocol) address of those accessing it. Officers will also work undercover, entering forums or chats posing as offenders, for example, to lure out the criminals acting on the Dark Web (Haasz, 2016). Hacking is also utilised by law enforcement in an attempt to catch illegal operators by reversing the anonymous status given to those on the Dark Web (Ghappour, 2017).

3 Challenges

Despite there being established methods undertaken by law enforcement to carry out digital policing, the Dark Web poses some serious concerns and challenges to this duty, which are to be detailed in this part of the study. As briefly covered in the previous section, the Dark Web can be used for both illicit or illegal, and legitimate or morally acceptable purposes, which raises the first dilemma for those digitally policing the Dark Web. The challenge for law enforcement is to be able to strike an appropriate balance between policing and putting a stop to the heinous crimes committed on the Dark Web, whilst also acknowledging the important role that the Dark Web plays in order to protect civil liberties, like freedom of expression and privacy, in a time where people are being monitored and controlled beyond what is necessary (Kumar & Rosenbach, 2019). The privacy offered by TOR can be utilised as a shield from persecution and a source of information for those living under oppressive regimes who are forbidden from accessing large parts of the Internet; in more liberal settings, it can be used as a vital whistleblowing apparatus allowing truths to be uncovered without fear of consequences (Ibid.). Contrastingly, the Dark Web provides a vast platform for illegal activities including a space to spread terrorist propaganda, the exchange of drugs, fake documents, illegal firearms and even the hiring of contracted killers (Naseem et al., 2016); however, research carried out by Owen and Savage (2015) suggests that the most popular content accessed on the Dark Web is relating to child abuse. With this in mind, the protection of political dissidents, whistle-blowers and those valuing their privacy should not cost the facilitation of child abusers, drug and firearm traffickers and other criminals (Kumar & Rosenbach, 2019). Therefore, action must be taken to find an equilibrium between the good and bad that is produced through the Dark Web.

Undercover operations led by law enforcement on the Web is not a new practice and involves the investigation of forums, chatrooms, newsgroups and various peer-to-peer networks in a covert manner (Haasz, 2016). Investigators will enter these spheres and pose as offenders or participants in illegal activities (Ibid.); in other circumstances law enforcement actors will take on the role of administrators of forums as a way of monitoring the activity without alerting the regular users (Staley & Montasari, 2021). This method has been used to tackle a plethora of different criminal activities, one being child abusers in which the undercover officers pose as children to lure in paedophiles (Ibid.). Although one of the most notorious cases being the arrest of Ross Ulbricht, the creator of Silk Road, which was one of the largest marketplaces for illicit goods on the Internet at the time (Frizell, 2015). A Department of Homeland Security investigator was able to take control of a lead website administrator's account and continued a line of communication with 'Dread Pirate Roberts', the anonymous creator of Silk Road, 'aiding' in the management of forums; soon after, Ulbricht was linked to the pseudonym and eventually arrested (Ibid.). Whilst there is empirical support for the usefulness of undercover operations, in recent times criminals are becoming more aware of the presence of such investigators on the Dark Web, thus there is growing suspicion and it is becoming increasingly challenging for undercover agents to come across as incontestable (Haasz, 2016). It is also important to note that whilst it was considered a success that Ulbricht was prosecuted for his crimes, replacements for the Silk Road marketplace were quick to reappear and were generally more difficult to shut down, leaving law enforcement back in the same position whereby they have no choice but to perform reactive online policing as opposed to preventative (Staley & Montasari, 2021). This consequently undermines the effectiveness of undercover operations and hinders these interventions becoming a long term and definite solution (Ibid.).

Another issue faced by law enforcement whilst policing the Dark Web is regarding their use of 'honeypot traps' and how ethical they are deemed to be. As previously mentioned, these traps are sites that have been set up by police containing illicit content which are intended to document the details of their IP address (Haasz, 2016). This method was utilised in an operation by the FBI in 2015 whereby a site called 'Playpen' was controlled containing links to 23,000 indecent images and videos of children, over 9000 of which were downloadable straight from the government's computer (Heath, 2016). Whilst the FBI was in control of the website it received over 100,000 visitors, by using a type of malware that revealed the IP addresses of users they were able to be charged (Ibid.). This type of operation raises much controversy as the FBI did not state that whilst the site was being monitored, there was no effort to prevent new images being uploaded or any way to cease the circulation and spread of the damaging material, essentially continuing the vicious circle of harm that the victims face. It has also been raised that there is a lack of empirical support for the effectiveness of tools such as honeypot traps, especially in terms of the cost and benefit analysis which still pays regard to the fact that it can be considered useful to an extent (Davis, 2017). However, taking account of the lack of overwhelming success and the serious ethical issues raised by this method, it suggests that other methods may be more appropriate.

Despite being a method associated with online criminals, law enforcement has also been able to make use of hacking to expose actors on the Dark Web, particularly as a way of stripping the users of their anonymity gained through TOR (Davis, 2017). This hidden nature is supported by the uncontrolled payment method through cryptocurrencies; specifically Bitcoin, a currency developed in 2011, the first of which that is not controlled by a national government and is therefore virtually untraceable (Kumar & Rosenbach, 2019). Hacking involves the remote installation of malware on a computer without gaining permission from the owner, finding a solution to the hidden location of users (Ghappour, 2017). Malware has the ability to upload files from one computer to a different server without the owner knowing and can also collect information that is processed by the microphone and camera on the target computer (Ibid.). Despite this being a very useful tool for policing the Dark Web, hacking does have its limitations, particularly when criminals are located in a different jurisdiction than the law enforcement (Jardine, 2015). This is due to a lack of consistency between laws across nations; and when a crime crosses between territories, the responsibility for law enforcement becomes shared and can make it difficult to administer justice in a fair way (Finklea, 2017). For example, the policy and legislation surrounding the use of TOR in the United States is very different and inconsistent with agendas displayed by China and Russia who are strongly opposed to the use of TOR in any circumstance (Chertoff, 2017). If criminals are to be pursued across physical borders, it can result in threats to national sovereignty that must be resolved efficiently.

4 Recommendations

Whilst the aforementioned challenges that the Dark Web poses to digital policing are significant, there are some recommendations that are able to counteract the issues faced. These suggestions are directed specifically towards law enforcement agencies and those with a research focus specifically in this sphere. First and foremost, it could be suggested that there are improvements made to the capabilities of domestic law enforcement, particularly in regard to the training of command level officers that are not just 'front line' investigators (Goodison et al., 2019). With further investments into training, it could ensure that the staff at junior levels are competent and have sufficient knowledge in all areas, and that the staff working at more senior levels are constantly reviewing the training to maintain a well-rounded and multifaceted curriculum (Goodison et al., 2019). This will go on to strengthen subsequent operations undertaken to intercept crime in the future, for example, it has been reported that the FBI has gained the ability to de-anonymise TOR servers, revealing the identities and locations of illegal websites residing on the TOR network (Kumar & Rosenbach, 2019). This can help to begin to take away the element that helps crime to thrive on the Dark Web, whilst allowing the legitimate sites to continue to run, leaving legal users with their privacy intact.

Another recommendation is that there should be new regulations implemented to de-anonymise and regulate the transactions of cryptocurrencies like Bitcoin that fund the illicit marketplaces on the Dark Web. This is following advice produced by leaders at the G20 Summit in 2018, in which it was suggested that international regulatory agencies were asked to review policy responses for crypto assets specifically in relation to combatting the financing of terrorism and anti-money laundering (Kumar & Rosenbach, 2019). One solution offered is by using a type of technology called Regulated and Sovereign Backed Cryptocurrencies (RSBC) which begins to de-anonymise the blockchain, a way of documenting the information about the sending and receiving of Bitcoins in a way that makes it very difficult to alter or access (Tewari, 2020). To do this, crypto coins should be converted into Nation Coins (the crypto currency of the nation), which allows information to be accessed about the person the crypto wallet belongs to and where the crypto coins are being spent; allowing trade for illicit material to be tracked (Ibid.). Even though crypto payment facilitators lack the required infrastructure to assume such financial industry measures, it is important to be proactive in laying the foundations for such surveillance to digitally police the Dark Web (Kumar & Rosenbach, 2019).

A fundamental third recommendation is to enhance the investment of money, resources and time into the communication across international borders between law enforcement agencies, making information sharing a more efficient process (Goodison et al., 2019). To improve cooperation, it would be helpful for nations to implement some mutually agreed upon legislation to police the Dark Web to ensure that there is common ground when international collaboration is needed to tackle certain cases (Chertoff, 2017). In turn, this will reduce confusion that has the potential to hinder investigations and compromise legitimate prosecution (Davis & Arrigo, 2021); key and useful bodies in these types of reviews and reforms would include Interpol and Europol (Hadjimatheou, 2017). Recent successful demonstrations of such methods showcase the potential, for example, following Operation DisrupTor, a series of connected operations planned by Europol and Eurojust to combat the trade of illicit substances on the Dark Web; there were 179 arrests across Europe and the United States (Europol, 2020). This illustrated how effective digitally policing the Dark Web can be when international agencies collaborate and have a shared goal.

5 Discussion

The key elements to note from this study are that the Dark Web is not inherently used for unethical and illegal behaviour, it can also be used as a tool to protect civil liberties around the world and assert a right to privacy. However, the anonymous nature it affords its users allows it to be exploited for a wide range of crimes including the trade of illicit substances, firearms, sharing of illegal content of child abuse and the hacking for financial gain, amongst others. This consequently poses new challenges for those aiming to digitally police the Dark Web, due to methods used like undercover operations, honeypot traps and hacking either becoming outdated and futile or raising ethical dilemmas; leaving the difficult task of trying to find a nuanced approach of protecting the advantages of an anonymised network whilst putting a stop to the criminal behaviours being enacted in liberal societies (Kumar & Rosenbach, 2019). The recommendations that have stemmed from this emphasise the acceptance of law enforcements inability to abolish the Dark Web itself and the following need to work on and improve methods of policing that are already utilised to increase the capabilities of forces (Davis, 2017). At the same time, by trying to implement new and safer regulations for payments being made via the Dark Web, and that this, amongst other shared goals are regulated and enforced in a similar manner across international borders (Chertoff, 2017).

Chertoff (2017) exemplifies the difficulties in differing stances on TOR and the Dark Web in the case of China, Russia and Austria all making efforts to de-anonymize or block TOR in different ways; and with the United States and Germany supporting it and going as far to governmentally fund it. The potential for success resulting from international cooperation is evident in previous cases, like in 2019 where combined international efforts were able to shut down 50 illegal websites on the Dark Web, two of which were Wall Street Market and Valhalla, notorious for being the largest dark marketplaces (Kumar & Rosenbach, 2019). Whilst this study offers some of the key challenges in digitally policing the Dark Web and ways to resolve this, the points are brief and this research would benefit coming from a more well-rounded perspective incorporating more computer science and cyber security knowledge, giving a more specified direction for improvement. Whilst there is lots of research available on the Dark Web, it would be beneficial if further research was specifically directed to aid law enforcement efforts and come up with detailed plans and methods that could be implemented.

6 Conclusion

This chapter began by offering a host of key background knowledge as to what the Dark Web is and its brief history, how it can be utilised, and how it can be digitally policed. This set the scene for the aim of this chapter and its research focus of discussing some of the challenges the Dark Web poses to digital policing, and the following objective of offering recommendations that may resolve these problems. Arguably the most relevant of which were highlighted in the discussion section, namely, the biggest challenge faced is finding an appropriate balance between enabling the positives of the Dark Web to be accessible whilst curbing the facilitation of illicit activities; additionally, the most effective recommendation if put in place is the power of shared international goals if enforced equally. Within the existing body of knowledge this chapter serves as a starting point to begin to understand how the Dark Web can impact digital policing and to start to think about ways in which digital policing can be kept up to date and progressive.

References

Bergman, M. K. (2000). *The deep web: Surfacing hidden value*. Bright Planet. http://resources.mpi-inf.mpg.de/d5/teaching/ws01_02/proseminarliteratur/deepwebwhitepaper.pdf

Bright Planet. (2014, March 27). *Clearing up confusion – Deep Web vs. Dark Web*.https://brightplanet.com/2014/03/27/clearing-confusion-deep-web-vs-dark-web/

Cabinet Office. (2011). *The UK Cyber Security Strategy: Protecting and promoting the UK in a digital world*.https://assets.publishing.service.gov.uk/government/uploads/system/uploads/attachment_data/file/60961/uk-cyber-security-strategy-final.pdf

Chertoff, M. (2017). A public policy perspective of the Dark Web. *Journal of Cyber Policy, 2*(1), 26–38. https://doi.org/10.1080/23738871.2017.1298643

Chertoff, M., & Simon, T. (2015). *The impact of the dark web on internet governance and cyber security. Global Commission on Internet Governance, 6*. CIGI/Canadian Electronic Library. https://www.cigionline.org/sites/default/files/gcig_paper_no6.pdf

Cyware. (2019, January 30). *How is surface web intelligence different from dark web intelligence?*https://cyware.com/educational-guides/cyber-threat-intelligence/how-is-surface-web-intelligence-different-from-dark-web-intelligence-393c

Davis, C. (2017, December 11). *Addressing the challenges of enforcing the law on the dark web*. The University of Utah. https://law.utah.edu/addressing-the-challenges-of-enforcing-the-law-on-the-dark-web/#_ftn8

Davis, S., & Arrigo, B. (2021). The Dark Web and anonymizing technologies: Legal pitfalls, ethical prospects, and policy directions from radical criminology. *Crime, Law and Social Change, 76*, 367–386. https://doi.org/10.1007/s10611-021-09972-z

Europol. (2020, October 1). *International sting against dark web venders leads to 179 arrests*.https://www.europol.europa.eu/media-press/newsroom/news/international-sting-against-dark-web-vendors-leads-to-179-arrests

Finklea, K. (2017). *Dark web*. Congressional Research Service. https://sgp.fas.org/crs/misc/R44101.pdf

Frizell, S. (2015, January 21). How the feds nabbed alleged silk road drug kingpin 'Dread Pirate Roberts'. *Time*. https://time.com/3673321/silk-road-dread-pirate-roberts/

Ghappour, A. (2017). Searching places unknown: Law enforcement jurisdiction on the dark web. *Stanford Law Review, 69*, 1075–1136. https://doi.org/10.2139/ssrn.2742706

Goodison, S. E., Woods, D., Barnum, J. D., Kemerer, A. R., & Jackson, B. A. (2019). *Identifying law enforcement needs for conducting criminal investigations involving evidence on the dark web*. RAND Corporation. https://www.rand.org/content/dam/rand/pubs/research_reports/RR2700/RR2704/RAND_RR2704.pdf

Haasz, A. (2016). Underneath it all: Policing international child pornography on the dark web. *Syracuse Journal of International Law and Commerce, 43*(2), 353–380. https://heinonline.org/HOL/Page? collection=journals&handle=hein.journals/sjilc43&id=361&men_tab=srchresults

Hadjimatheou, K. (2017). *Policing the dark web: Ethical and legal issues*. University of Warwick. https://ec.europa.eu/research/participants/documents/downloadPublic? documentIds=080166e5c2573eef&appId=PPGMS

Heath, B. (2016). FBI ran website sharing thousands of child porn images. *USA Today*. https://eu.usatoday.com/story/news/2016/01/21/fbi-ran-website-sharing-thousands-child-porn-images/79108346/

Jacoby, C. (2016). *The onion router and the Darkweb*. Tufts University. https://www.cs.tufts.edu/comp/116/archive/fall2016/cjacoby.pdf

Jardine, E. (2015). *The dark web dilemma: Tor, anonymity and online policing. Global Commission on Internet Governance Paper Series, no. 21*. CIGI/Canadian Electronic Library.

Kaur, S., & Randhawa, S. (2020). Dark web: A web of crimes. *Wireless Personal Communications, 112*, 2131–2158. https://doi.org/10.1007/s11277-020-07143-2

Kavallieros, D., Myttas, D., Kermitsis, A., Lissaris, E., Giataganas, G., & Darra, E. (2021). Using the dark web. In B. Akhgar, M. Gercke, S. Vrochidis, & H. Gibson (Eds.), *Dark web investigation* (pp. 27–48). Springer Nature Switzerland AG). https://doi.org/10.1007/978-3-030-55343-2_2

Kumar, A., & Rosenbach, E. (2019). The truth about the dark web. *International Monetary Fund, 56*(3), 1–4. https://www.elibrary.imf.org/view/journals/022/0056/003/article-A007-en.xml

Kushner, D. (2015, October 22). The Darknet: Is the government destroying 'the wild west of the internet?'. *Rolling Stone*. https://www.rollingstone.com/politics/politics-news/the-darknet-is-the-government-destroying-the-wild-west-of-the-internet-198271/

McCormick, T. (2013, December 9). The Darknet*: A short history. Foreign Policy*. https://foreignpolicy.com/2013/12/09/the-darknet-a-short-history/

Moore, D., & Rid, T. (2016). Cryptopolitik and the Darknet. *Global Politics and Strategy, 58*(1), 7–38. https://doi.org/10.1080/00396338.2016.1142085

Naseem, I., Kashyap, A. K., & Mandloi, D. (2016). Exploring anonymous depths of invisible web and the digi-underworld. *International Journal of Computer Applications, 3*, 21–24. https://research.ijcaonline.org/nccc2016/number3/nc3-2016356.pdf

Ning, H., Ye, X., Bouras, M. A., Wei, D., & Daneshmand, M. (2018). General cyberspace: Cyberspace and cyber-enabled spaces. *IEEE Internet of Things Journal, 5*(3), 1843–1856. https://doi.org/10.1109/JIOT.2018.2815535

Owen, G., & Savage, N. (2015). *The Tor dark net. Global Commission on Internet Governance, no. 20*. CIGI/Canadian Electronic Library. https://www.cigionline.org/sites/default/files/no20_0.pdf

Staley, B., & Montasari, R. (2021). A survey of challenges posed by the dark web. In R. Montasari & H. Jahankhani (Eds.), *Artificial intelligence in cyber security: Impact and implications* (pp. 203–213). Springer. https://doi.org/10.1007/978-3-030-88040-8_8

Sui, D., Caverlee, J., & Rudesill, D. (2015). *The deep web and the darknet: A look inside the internet's massive black box*. Wilson Center. https://www.wilsoncenter.org/sites/default/files/media/documents/publication/deep_web_report_october_2015.pdf

Tewari, S. H. (2020). *Abuses of cryptocurrency in dark web and ways to regulate them*. Available at SSRN 3794374

Zhang, H. G., Han, W. B., Lai, X. J., Lin, D. D., Ma, J. F., & Li, J. H. (2015). Survey on cyberspace security. *Science China Information Sciences, 58*(11), 1–43. https://doi.org/10.1007/s11432-015-5433-4

Pre-emptive Policing: Can Technology be the Answer to Solving London's Knife Crime Epidemic?

Sandra Smart-Akande, Joel Pinney, Chaminda Hewage, Imtiaz Khan, and Thanuja Mallikarachchi

Abstract As knife-related crimes continue to increase it has become an ever-growing area of concern in the UK. The highest rates were recorded in the London Metropolitan area with a rate of 168 offences involving a knife per 100,000 people in 2017/2018. This is an increase of 26 offences per 100,000 people since the previous year's statistics (2016/2017). With knife-related incidents continuing to climb on an upwards trend, the Metropolitan Police continues to explore innovative methods to address these worrying statistics. This chapter provides an in-depth analysis on the research of knife-enabled crimes, causes and motivating factors with the focus on the 33 boroughs of London. This chapter reviews the contemporary literature to answer many of the important questions including what are the main causes and who are perpetrators of knife crimes? how are the London Metropolitan Police dealing with knife crime? and how does the London Metropolitan Police force employ big data and predictive analysis for pre-emptive policing? The findings of this chapter uncover not only an analysis of the recent knife crime statistics in London but also a review of the motivations causing individuals to carry a bladed article. The chapter also addresses how technological innovations can be utilised to address the knife crime epidemic. Through investigating pre-emptive policing using big data and innovative technology, the chapter provides a critical review on how technology can help the London Metropolitan Police address knife crime.

Keywords Knife crime · Pre-emptive policing · Big data · Predictive analysis

1 Introduction

Knife crime is an epidemic that is growing consistently in the UK. This is concerning because it disproportionately impacts the young, the venerable or

S. Smart-Akande (✉) · J. Pinney · C. Hewage · I. Khan · T. Mallikarachchi
Cardiff Metropolitan University, Cardiff, UK
e-mail: ssmart-akande@cardiffmet.ac.uk; jpinney2@cardiffmet.ac.uk;
chewage@cardiffmet.ac.uk; IKhan@cardiffmet.ac.uk; TMallikarachchi@cardiffmet.ac.uk

R. Montasari (ed.), *Artificial Intelligence and National Security*,
https://doi.org/10.1007/978-3-031-06709-9_11

disadvantaged individuals. It is more predominant in the ethnic minority groups (Policy Exchange 2021). Not all Londoners are affected equally by knife crime, there is a large discrepancy across borough of residence, ethnicity, age and gender (Policy Exchange 2021). In this chapter, we define young people as Londoners between the ages of 10 and 25 years of age. Offences involving young people with knife and gun crimes have been at the forefront of media and public attention for many years (Arianna et al. 2009) and a constant focus for the London Mayor Office (Mayor of London 2021). Tackling knife crime is a priority for the London Metropolitan Police Force (Metropolitan Police 2022) and the UK government is determined to do everything to break the effect of crime and violence that is devastating lives in the communities and families.

In the year ending March 2021, the Office of National statistics (ONS) reported approximately 41,000 selected offences involving a knife or sharp instrument in England and Wales (Allen et al. 2021).

London had the highest rate of 168 knife-related offences per 100,000 population in 2017/2018, this was an increase of 26 offences per 100,000 population from 2016/17. The lowest rate was recorded in Surrey with 5 offences per 100,000 population (up by 1 from 2016/17) (Allen et al. 2021).

Surveys and statistics have shown that young people not only have been responsible but also make up the majority of victims of knife- enabled crime, children as young as 13 years old, and have been victims and perpetrators of the crime. Studies have shown evidence to point to the fact that the vulnerable population made up of looked-after-children, children in foster care, homeless young adults and children excluded from school have increased since 2014 as the levels of knife crime have increased (Ford 2018). In a survey of young people aged up to 25 years conducted in England and Wales from 2003–2006 by the Centre for Crime and Justice Studies, they reported that knife carrying was associated with being male, having a past record for criminal offence, drug use, violent bullying or victimisation, having delinquent friends and showing a lack of trust for the police (Ford 2018).

1.1 Statistics

The Metropolitan Police Service (MPS) Crime statistics 2020/21 dashboard (Met Police Data Board 2021) records the number of crimes involving a knife or sharp instrument on their Met Police crime data dashboard. In this database, the incidence of all crime committed in every London Borough is recorded and updated by the MPS. Homicides are grouped into six method of Killings: Knife or sharp instrument, blunt instrument, physical assault without weapon, other method of killing, not known or recorded, burning or scalding, other poisoning (drugs, etc.) and shooting. Knife attacks in London are heavily concentrated in certain boroughs. The worst affected boroughs are Lambeth, Newham, Southwark, Brent, Hackney and Haringey. Kingston, Sutton and Richmond upon Thames were reported to have

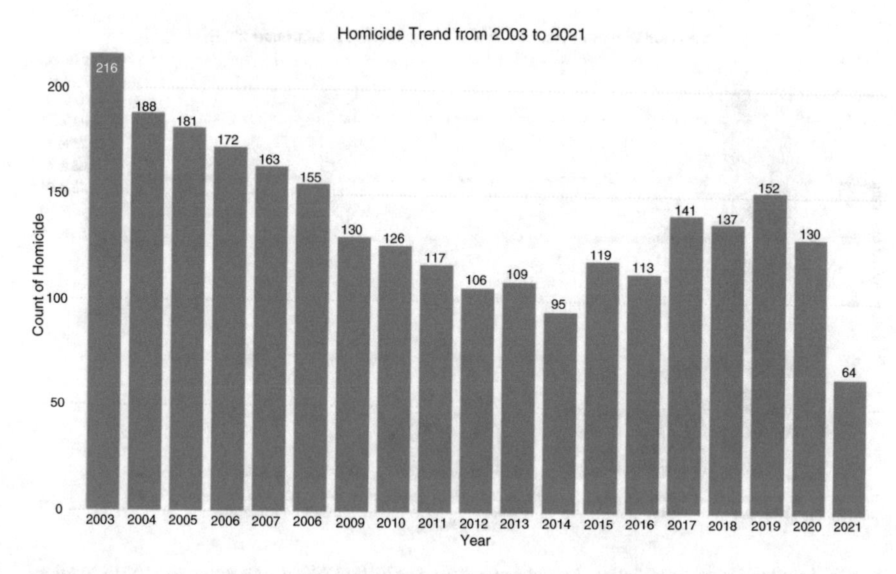

Fig. 1 London Homicide Victims 2003–2021. Source: MPS homicide dashboard data. Accessed February 2022

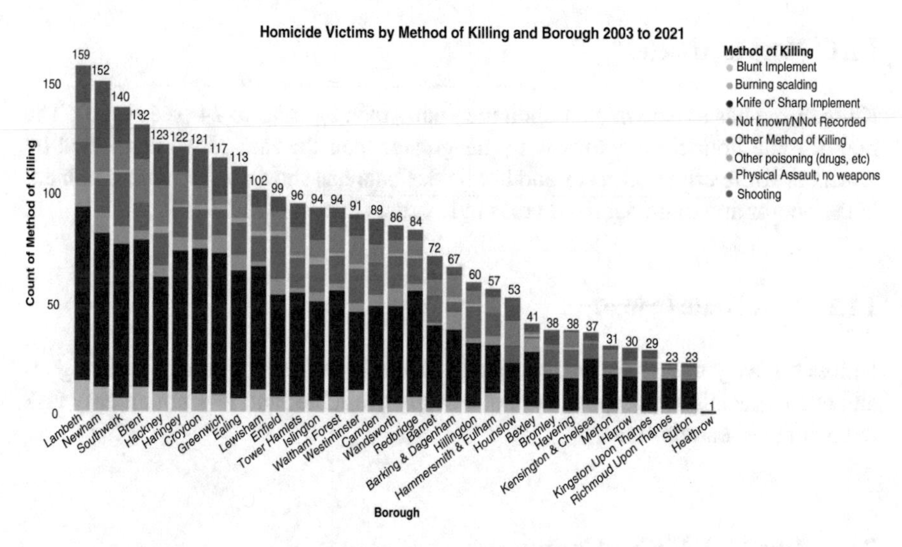

Fig. 2 London Homicide Victims by method of killing and borough 2003 to 2021. Source: MPS homicide dashboard data. Accessed February 2022

the lowest count. Figure 1 shows the total count of homicides across the London boroughs from 2003 to 2021; Fig. 2 depicts the count of homicides across the 33 boroughs of London showing the different categories of homicides committed per year.

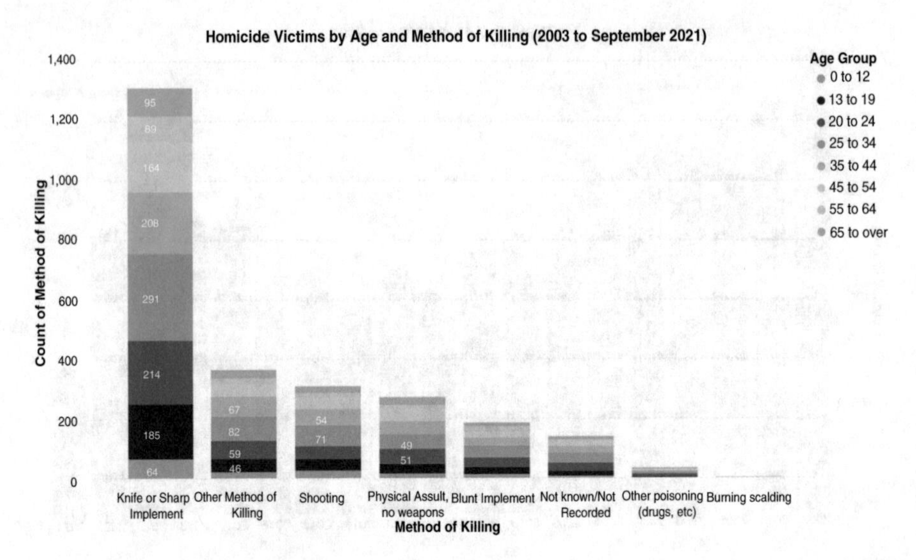

Fig. 3 London Homicide Victims by age and method of killing (2003 to September 2021). Source: MPS homicide dashboard data. Accessed February 2022

1.1.1 Who Is Affected?

Knife crime has a huge impact upon the young men aged 13 to 24 (see Fig. 3). The rate of male homicide victims is by far greater than the rate of women killed by violence. Knife crime offences and homicides data has shown to be more dominant in the poorer and more deprived areas in London (Brown et al. 2019).

1.1.2 Why Knife Crime?

Knives are easy to acquire and readily available in almost any home. Whilst guns and other specialised weapons are not 'readily available' and it is not an easy task to obtain live ammunition.

2 What Is A Knife Crime

Violent crime is defined as any action that intentionally inflicts or threatens a person's physical or causes psychological damage (Houses of Parliament 2019). Knife crime is defined as any criminal activity that involves an object having a blade or sharp instrument (Allen et al. 2019). This can be used to commit a range of offences including

- Robbery
- Violence against the person
- Burglary
- Sexual offences and domestic violence
- Criminal damage
- Enforcing authority (Arianna et al. 2009).

The law in the UK states that 'it is an offence under Sect. 1 of the Prevention of Crime Act 1953 for a person to have with him in any public place any offensive weapon without lawful authority or reasonable excuse'. Offensive weapon in Sect. 1 of the Criminal Justice Act is defined as: 'any article made or adapted for use for causing injury to the person, or intended by the person having it with him for such use by him or some other person' (Wheller 2019; CPS 2021). Under section 139 of the Criminal Justice Act 1988, it is an offence for a person to carry in a public place any article which has a blade or is sharply pointed, except a folding pocketknife with a cutting edge of three inches or less, without good reason or lawful authority. The Crown Prosecution Service (CPS) (CPS 2021) states three categories of weapons:

1. 'A weapon made for causing injury to a person—'offensive per se'. Examples of these knives include a swordstick, flick knives, shurikens (or 'death stars'), butterfly knives, a handclaw, a belt buckle knife, a push dagger, a 'kyoketsu shoge', being a length of rope, cord, wire or chain fastened at one end to a hooked knife. Weapons which are manufactured for an innocent purpose, such as a razor, a penknife and some types of sheath knives, are not offensive per se.
2. A weapon not made for that purpose but adapted for it, e.g. objects with a razor blade inserted or mounted into them or a deliberately broken bottle.
3. A weapon neither made nor adapted to injure but one which is intended by the person to be used to injure, such as where a defendant has with him a lock knife, a Stanley knife or a dagger with intent to injure. The fact that a person is carrying an object which they go on to use to injure another is not necessarily evidence of intent; a person may lawfully carry a penknife and then use it to injure another but at the time they were carrying it, it was not an offensive weapon. This will be inferred from the defendant's actions and the circumstances surrounding the possession. For example, following a statement to the police that the weapon was carried for self-defence, an inference could be drawn that for the purposes of defending himself a defendant would, if necessary, use the knife to cause injury. However, an intention to frighten is not enough to satisfy this, unless it is so intimidating as to be enough to produce an injury through the operation of shock'.

3 Motivations and Risk Factors for Carrying Knives

Why People Carry Knives?

The College of Policing Evidence Briefing (Wheller 2019) identified three broad categories of why people carry knives

I. Self-protection and Fear ('defensive weapon carrying'): This is a defence strategy particularly for individuals who have suffered as a victim of crime.
II. Self-Presentation: This is mainly for individuals who want to show off by having 'street credibility' and 'respect'.
III. Utility (offensive weapon carrying): This is particularly for individuals who want to facilitate criminal behaviours and intimidate others with theft, sexual assault, injury and serious harm.

Four Typology of Young People that Carry Weapons and Knives (Lemos 2004):

Group A. Young people who have past offence record and are in the criminal justice system.

Group B. Associates of offenders—this includes those in group A mentioned above—these have not been identified by the criminal justice system and, therefore, have no involvement with probation or government agencies.

Group C. This group is made of young people who carry knives most of the time and are already known to the youth, education and criminal agencies.

Group D. These are young people that carry weapons only when they feel that there are known risks and carry these weapons without the knowledge of any agencies.

There is no individual factor that causes a young person to become a perpetrator or victim of violence. A combination of risk factors can accumulate to create experiences early in a person's life that leaves the individual venerable to exploitation or crime (Houses of Parliament 2019). Many researchers, groups, agencies, police and government organisations have proposed ways of reducing the crime rates. Factors and risks of an individual being a victim or a perpetrator are constantly being identified and the more common ones are listed in this chapter.
It should be noted that there is no straightforward inter-dependencies behind these risk factors but they are likely to interact with each other and with other risk factors. The direction of causality may be vague with some risk factors.

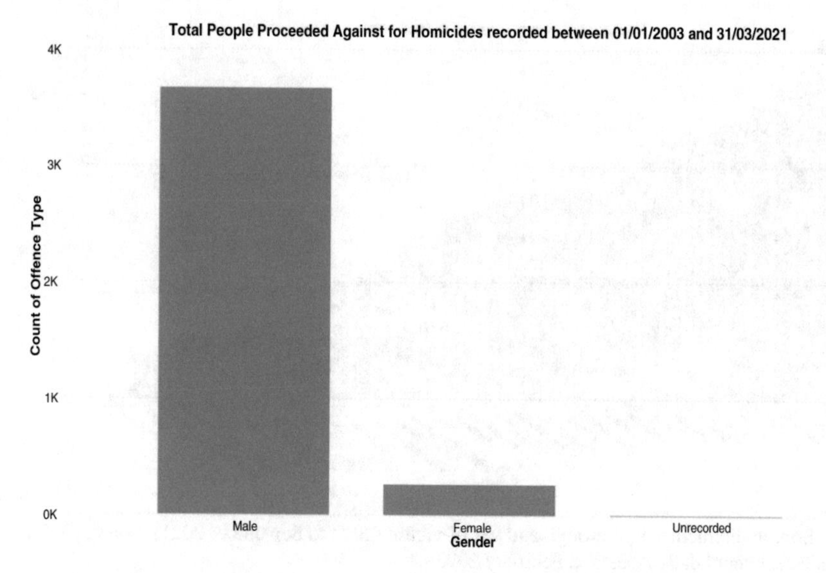

Fig. 4 Total people proceeded against for homicides in London recorded between 01/01/2003 and 31/03/2021. Source: MPS homicide dashboard data. Accessed February 2022

3.1 Risk Factors

3.1.1 Gender

According to research males and females carry knifes but the males generally outnumber the females (Lemos 2004). Gangs are mainly males (Marshall et al. 2005). Serious violence and carrying of weapons are likely to be committed by males (Wheller 2019). Figure 4 shows a high proportion of males as the total people proceeded against for homicides recorded between 01/01/2003 and 31/03/2021.

3.1.2 Ethnicity

Black, Asian and Minority ethnic (BAME) groups are mainly affected in crimes involving knives/sharp instruments (Houses of Parliament 2019). The College of Policing suggests that there is no evidence to prove statistically significant relationship between ethnicity and weapon carrying (Wheller 2019).

3.1.3 Age

Statistics have shown the peak age for young people carrying knives and other weapons to be around 15, 16 years old (Henke et al. 2016; Al-Kassab et al. 2014).

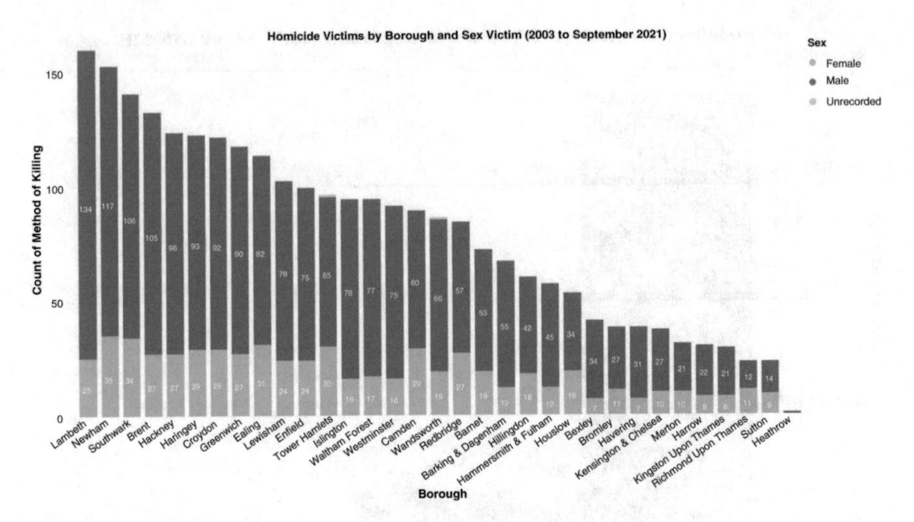

Fig. 5 London homicides by borough and sex of victim (2003 to September 2021). Source: MPS homicide dashboard data. Accessed February 2022

Literature suggest that most gang members are under the age of 18 years and the oldest age being around 25. Young people around 9 years old can become involved in street gangs but the majority of the gang-prone kids become involved around the ages of 12 and 14 (Marshall et al. 2005). Figures 5 and 6 show the large discrepancies between the females and males both as victims and perpetrators. The National Crime Agency Intelligence Assessment reports that individuals of this age groups are likely targeted as older criminals find them easier to use, manipulate, control, exploit and reward than adults (National Crime Agency 2019).

3.1.4 School Exclusion/Social Exclusion

Young people that have been excluded from mainstream education have been known for carrying knifes; this has been a common occurrence in the past years. The College of Policing suggests that educational attainment and school exclusion are a risk factor associated with knife crime (Wheller 2019). They highlighted that school exclusion and lack of adequate alternative education can cause issues of low achievement thereby creating a disadvantage in the labour market and further increasing the discernment of lack of opportunities; this further leads the individual to a life of engaging in crimes. There is a risk that young people that have been rejected from institutions will get involved in gangs where delinquency is celebrated and problematic behaviour is thought to be normal (Marshall et al. 2005).

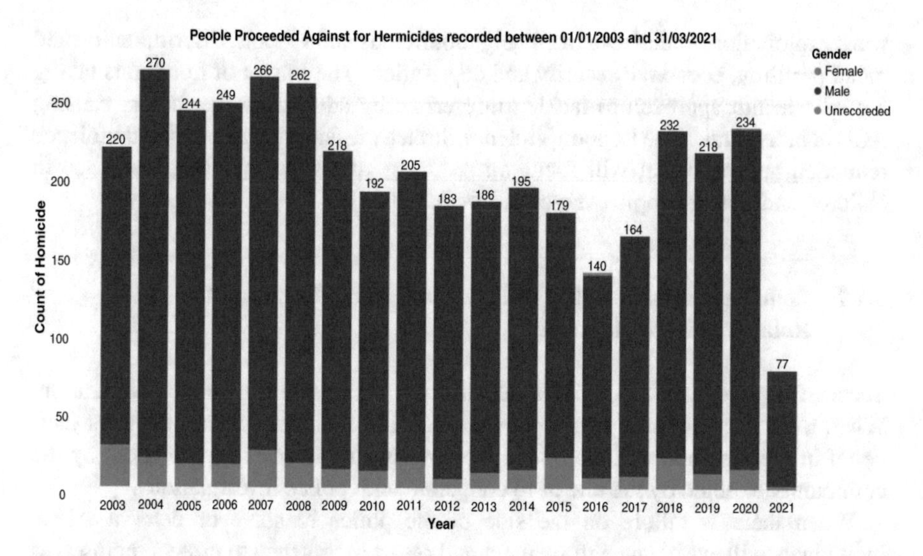

Fig. 6 People proceeded against for homicides in London boroughs recorded between 01/01/2003 and 31/03/2021. Source: MPS homicide dashboard data. Accessed February 2022

3.1.5 Gangs

Gangs are formed through processes of discrimination and/or social exclusion when these young people come together to protect themselves feeling that there is 'safety in numbers'. Immigrant population and similar groups are likely to be particularly at risk as they also can find solace in their similarities and perceived social exclusions (Marshall et al. 2005).

3.1.6 Adverse Childhood Experiences (ACEs)

Studies have suggested ACEs are being a direct cause of knife crime (Mayor of London 2021). Adverse childhood experiences are occurrences in the early life of an individual encompassing of various forms of emotional and physical abuse, child maltreatment, neglect, exposure to family alcohol misuse, unstable family life and any experience that would leave a child with an increased risk of negative health or social outcomes in the latter life of the individual. Multiple ACEs in the life of an individual have been associated with poor mental health and/or self-harming or violence towards another person (Houses of Parliament 2019). A study in the USA (Barajas-Gonzalez et al. 2021) highlights ACE in children brought up by immigrant families and suffering as a result of being directly exposed to the experience of detention and deportation or threats of detention and deportation. This has shown to be a direct cause of adverse experiences for many immigrant children. Other immigration-related ACEs given are: Precluded access to resources, parental work exploitation, under-resourced neighbourhoods and schools, discrimination and racial profiling, economic security and deprivation. The Mayor of London is taking a public health approach to tackle knife crime by addressing the factors causing ACE in individuals. The London Violence Strategy is adopting a contextual violence reduction approach that will focus on reducing ACEs and building resilience in children and young people (Mayor of London 2021).

3.1.7 Poor Relationship with the Police and Local Community Relationships with Police

Successful police work in an area depends profoundly on the public cooperation. When a local community has a poor relationship with the police, it lowers their belief in the legitimacy of the police thereby causing a lack of willingness by the community to abide by the law or to cooperate with police investigations.

When there is failure on the side of the police to solve or deter a crime, individuals will evade law enforcement and resort to violence to protect themselves or resolve disputes.

This lack of trust in the authorities also causes a large portion of crime in a community to go unreported; this can be also due to a fear of retribution for reporting to the authorities (Marshall et al. 2005).

3.1.8 Children and Young People Do Not Trust the Authorities to Protect Them

A young person already involved in crime and living in a high-crime neighbourhood may not see the police as willing or being able to protect them from danger. This situation will make the individual feel justified for carrying weapons (Brennan 2018). Young people are reported to have had preconceived ideas and the perception of unfair treatment in police tactics and have felt unfairly monitored by the authorities (Houses of Parliament 2019). These young persons bypass law enforcement and do not seek or rely on police procedures for protection; this lack of trust in the police has been a leading cause of victims becoming perpetrators of crime as they aim to seek revenge or protection for them instead of relying on police procedures (Wheller 2019; Brown et al. 2019).

3.1.9 Poor Mental Health

Adverse childhood experiences have been a leading cause of metal health problems in individuals and a poor mental health has been identified as a factor that increases the likelihood of violent behaviour in an individual (Ford 2018; Houses of Parliament 2019). The Chief Medical Office for England reported in 2012 that around 10% of children and young people are affected by poor mental health (Houses of Parliament 2019).

3.1.10 Family Life

Offenders target children from seemingly stable backgrounds and exploit their vulnerabilities such as problems with parents or with their peer group. Children that have no criminal footprints are also a lucrative target for these criminal individuals who attempt to use them to reduce or deter attention from law enforcement (National Crime Agency 2019).

Children living with the absence of a suitable male role model in a home, such as a responsible father, are lured into joining gangs; this gives them the feeling of inclusion and belonging to a family and resulting in seeing the older criminal or gang leader as a brother or a substitute father (Marshall et al. 2005).

Young persons displaying other vulnerabilities such as poverty, intervention by social services, family breakdown and looked after status are a frequent target for county line offenders (National Crime Agency 2019).

3.1.11 Peer Relationships and Friends

A consistent finding throughout literature is that individuals are more likely to be involved in crime or problem behaviour when they have relatives or peers with a history of criminal or problem behaviour. Individuals whose families and/or peers are not involved in guns, drugs or gangs are less likely to be offenders themselves.

3.1.12 Material Aspirations

County line offenders recruit victims that have limited economic opportunities and resources by flaunting payments and material gains that would not be easily obtainable through legal means. This is very common with the availability and convenience of social media, on these platforms, images and videos of expensive trainers, cash, designer wears, luxury cars and a lot of flashy and high value goods are posted, creating a fallacy that it is rewarding to be involved in crime (National Crime Agency 2019)

3.1.13 Geographical Location

Literature has shown that living in an area with reduced employment opportunities and low aspiration is a risk factor associated with crime (Houses of Parliament 2019). A trend that is emerging in drug recruitment is targeting young children within importing cities and towns rather than targeting children in the high crime exporting areas where the offending groups are based.

Other key locations targeted by offenders for recruitment include schools, further and higher education institutions that were previously attended by the offenders, foster homes, special education institutions, pupil referral units and homeless shelter (National Crime Agency 2019).

3.1.14 Poverty, Inequality, Deprivation and Austerity

The majority of knife crime takes place in disadvantaged neighbourhoods that are suffering from huge social disinvestment (Brown et al. 2019). A child displaying poverty as a vulnerability is an easy target for criminal recruitment and offending drug groups (National Crime Agency 2019). The Youth Violence Commission (Houses of Parliament 2019) has identified poverty to be a potential source of violence. Young people are lured by gangs into a world of thrills and excitement, with the promise of high status material goods like jewellery, mobile phones and trainers. Austerity has been linked to the rise in violent crime, in these same communities; there has been a reduction in police resourcing and funding, reductions in various government institutions including prisons, social care, probation services, mental and social health services as well as youth services and government services.

The lack of support, services and opportunities readily available to vulnerable young people living in deprived areas are key issues that are linked to knife crime in the UK (House of Lords 2019; Warrell 2018; Chakelian 2019).

3.1.15 Peer Pressure or Gang Involvement

A study by the Institute of Crime Science, University College London (Marshall et al. 2005), defined street gangs as a groups of individuals that share an identity, which may be centred around a particular location or ethnicity or share a common identity based on age, friends (peer networks) or blood relatives. In a gang community, younger gang members are used to store drugs and weapons, drive vehicles or act as a look out. These young gang members are compelled to compliance by fear and intimidation; they are afraid to provide evidence or corporate with law enforcement; they find it difficult or impossible to disengage from the group. Violence and knife crime have been committed in the past as a result of conflicts between organised crime groups and resulting in fatal outcomes.

3.1.16 Drug Dealing

The more organised drug trafficking criminal groups use the lower level street gangs for street and their peers for dealing on the streets. The strategy employed here is that these individuals are used for the more ‘risky and visible’ tasks. This helps the criminals that are higher up the chain to avoid unwanted police attention (Marshall et al. 2005; House of Lords 2019). Children and vulnerable adults are often targeted and used as drug runners to transport the drugs to other parts of the cities and neighbouring towns and to collect payment. This is because they are less likely to be suspected as being involved in drugs than adults (House of Lords 2019; National Crime Agency 2019).

Drug dealers seeing themselves as desirable targets for robbery will arm themselves with weapons such as knives and guns to protect themselves and territory and also use these weapons to enforce debts (Marshall et al. 2005; House of Lords 2019).

3.1.17 Drug Using

Involvement with drugs has been growing over recent years for young people; there has been an increase in the use of recreational drug (for example, cannabis) among those aged 11 to 15 years old. The conviction rate for young people aged 10 to 17 years for possession with intent to supply has increased by 77% between 2012 and 2016. This is triple the amount of adult convictions. Drug using has been the cause of a lot of serious mental health problems and a leading cause of serious violence and knife crime (House of Lords 2019).

3.1.18 Previous Arrest/Criminality

Children that have had previous involvement with criminality including drug offending are a lucrative target for drug offenders (House of Lords 2019).

3.1.19 Fear and Intimidation, Protection

Young persons carry weapons for a number of reasons; the most common reason is for protection. The College of Policing identified self-protection and fear ('defensive weapon carrying') as a motivation for carrying knives, mainly for individuals that have recovered or being a victim of crime (Wheller 2019; Marshall et al. 2005).

3.1.20 Social Status and Respect

Motivations for joining gang include respect and power (Wheller 2019). The College of Policing identified self-presentation as a motivation for carrying knives especially for individuals who want respect and credibility on the street or among his peers (Wheller 2019).

3.1.21 Utility

In utility weapon carrying, traffickers and drug dealers use firearms and bladed weapons to protect their territory and enforce debts (Marshall et al. 2005). Offensive weapon carrying is particularly for individuals who use weapons like firearms and knives to enforce behaviours such as sexual assault, theft, injury and serious harm.

3.1.22 Victim of Violent Crime

Studies have shown that it is five times more likely for young people who have been victims of violent crime or trauma to offend than other people. This is a result of the factor of fear and victimisation that causes a young person to feel the need to arm themselves with a knife in order to preserve safety (Marshall et al. 2005). Evidence indicates that those involved in perpetuating knife crimes have frequently been victims of violence themselves (Houses of Parliament 2019).

3.1.23 Bullying

Bullying in individuals has caused the individual to feel disposed to weapon carrying as a means of protection or dispelling timidity to the oppressor. Bullying means that both victims and perpetrators are more likely to carry a weapon than others.

3.1.24 Social Media

Knife crime has been linked to the rise of drill music which is a form of rap and mainly posted on social media platforms. Drill music is associated with lyrics that are abusive of the law enforcement, glamorising crime, glamorising serious violence, murder, stabbing and they go into details to describe these crimes with great joy and excitement using upbeat music.

The lyrics in drill music have been known to incite real-world violence among young people and have made the Metropolitan Police Force to request that the popular social media platform, YouTube, to remove some of the videos inciting violence. Gangs in London have been known to taunt each other using drill music to brag about crime they intend to commit or have committed (House of Lords 2019).

Some offenders have been known to use social media as a means of intimidating other rival offenders recruiting, promoting their identity and band, and also as a means of recruiting potential offenders.

4 The London Metropolitan Police Force

The mission of the Metropolitan Police Service (the Met) is to Keep London Safe for Everyone. According to Falkner et al., 'Effective policing with regards to knife crime requires a few key approaches, including suppression, targeting high profile criminals and gang members, and neighbourhood policing. These all contribute towards the vital task of bringing the perpetrators of knife crime to justice. None of these alone is sufficient to tackle knife crime, and in order to be effective the correct balance between these different measures must be struck in order to be effective' (Policy Exchange 2021). Sophie Linden, Deputy Mayor for Policing and Crime said in the Police and Crime Plan 2021–2025, 'While the MPS plays a fundamental role in preventing crime and keeping people safe, they cannot do this alone. It will require dedicated and sustained partnership working with individuals, communities, voluntary and civil society organisations, Community Safety Partnerships, the Crown Prosecution Service, the National Health Service, the government and its agencies, and many others. We are building on strong foundations but over the next few years we will deepen and strengthen these partnerships even further. The Mayor and I will also continue to lobby for increases in Government funding for preventative and youth services, and for greater long-term certainty on that funding to enable them to build further on their work' (Mayor of London 2021).

4.1 *Structure of MPS*

The Met is the largest police force in the UK and it is made up of more than 43,000 officers and staff. It has 25% of the total police budget for England and Wales (Metropolitan Police 2022).

As of 31 January 2022, the Met is made up of the following personnel divided up within its various departments:

Police officers	33,128
Police staff	9804
Police community support officers	1179
Special officers	1838

4.1.1 The Met Comprises Four Business Groups

Front-line Policing, Specialist Operations, Met Operations and Professionalism. These groups are supported by a single Met HQ that provides strategic services

covering People and Change, Commercial and Finance, Legal, Media and Communications, Strategy and Governance and Digital Policing.
Digital Policing is the Met's technology function. The Digital Policing Strategy 2021–2025 document sets out Digital Policing's strategy to helping realise the Met's digital vision 'Digital policing will make it easier for the public to make contact with the police wherever they are in the country, enable us to make better use of digital intelligence and evidence and transfer all material in a digital format to the criminal justice system' (NPCC 2025).

4.2 Policing Strategies in London Metropolitan Police Service (MPS)

4.2.1 Hot-Spot Policing

The UK Police force data reveals the hot-spots where there are high counts of crime in certain areas. These areas show where serious violence is concentrated and the strategy of the MPS is to implement a Problem Oriented Policing (POP) approach by prioritising resources to the areas that have higher risk of violence. The POP is the strategy of using intelligent analysis to deal with specific crime by understanding the issue and deploying tailored responses to tackle it and making an assessment to understand what has worked well and what can guide any changes needed (Mayor of London 2021).

4.2.2 Stop and Search

The police have the power to deploy a Sect. 1; this means that they can stop and search someone if they have 'reasonable grounds' to suspect they are in possession of drugs, a weapon, stolen goods or something that could be used to commit a crime.

A section 60 is deployed if the police believe a serious crime has been committed at a particular time in a particular location.

The Home Office had recently allowed the police to use section 60 in seven areas in London with the highest knife crime. Stop and search is more effective when it is intelligence lead, because research data has shown that the increased use of stop and search does not increase arrests, and random stop and search does not reduce the counts of violent crime in an area.

4.2.3 Enhanced Mobility

Digital Policing of the MET has made over 40,000 mobile devices available across its workforce. Front-line Officers are equipped with second generation ruggedized tablets that are robust and enable officers in the field to complete their work with minimal need to return to desks at the police stations. In addition, they will also be equipped with smartphones, thereby having the flexibility to receive intelligence on the move and offering more mobility with technology.

Detectives, specialists and back office staff in the MET are equipped with laptops. These laptops have enabled staff to work remotely and smartly (Metropolitan Police Service 2021, p.16).

5 Embracing Innovative Technology to Tackle Knife Crime

In today's society the ability to capture data is everywhere, from CCTV (Closed-circuit television) to the IOT (Internet of things), policing in the UK has become a technology driven force. The transition to becoming technological driven has been well received and named 'Digital Policing' in the Metropolitan Police force (Deloitte 2015). The purpose of digital policing will act as the Metropolitan police's function to ensure the commitment of utilising information technology (McCallum 2020). Moreover, managing their digital transformation of using high tech computing, body video capturing devices and providing easier access to digital data sources such as CCTV. Since the first permanent deployment of CCTV in London in the 1960's (Williams 2003), the UK has grown to be named the 'most surveyed' country with an ever-growing number of CCTV cameras installed (Webster 2009). It is said that the global CCTV camera coverage could surpass one-billion in 2022 and of which half a million cameras will be located in London alone (Sintonen et al. 2021). It is estimated that the average person living in London will be caught on a CCTV camera 300 times per day (Sheldon 2012). With the growing increase in CCTV coverage throughout London, the use of facial recognition technology and infrared cameras are becoming the widespread expectation for physical security. While CCTV cameras are valuable evidence for reviewing crimes and prosecution cases they have also been shown to offer a significant increase in the chance that a crime is solved (Ashby 2017). However, it is important to consider that not all crimes can be identified clearly with CCTV imaging alone and with the added factor CCTV may not cover the area of interest (Ashby 2017). Furthermore, when exploring crimes involving deadly weapons such as knifes, we must consider that knives are inherently small and easily concealable objects and without a clear line of sight will not be picked up with conventional CCTV camera coverage. However, advancements in technology have began exploring the use of automated detection systems. Whilst the problems surrounding a knifes shape, texture and size may still cause issues for automated detection, research by Buckchash and Raman propose solutions to address these issues (Buckchash and Raman 2017). Through their proposed approach of FAST (Features from Accelerated Segment Test), foreground segmentation and MRA (Multi-Resolution Analysis) they show promising results to develop robust object detection algorithms for the issues of knife detection. However, controversially another factor to consider is knifes are made of reflective metal/steel that may distort imaging and may not produce a clear representation of an individual carrying a bladed weapon (Galab et al. 2021). Although technology does exist to help minimise these effects (Galab et al. 2021), a further advancement in technology has now started exploring thermal imaging surveillance technology. Thermographic cameras or simply infrared cameras are used to detect body heat and can be used to discover hidden concealed weapons such as knives (Timberlake 2020). Moreover, thermal imaging technology allows for a non-intrusive examination of a person belongings that can penetrate through clothing to uncover the attempted concealed bladed article. The metropolitan police have stated their interest of trialling these thermal imaging scanners as a means to address knife crime (Davenport 2022). A Canadian company 'Patriot One Technologies' thrives on offering twenty-first century multi-sensor covert threat detection equipment to combat threats before they occur (PatriotOne 2022). Martin Cronin the CEO of Patriot One explained 'If an individual with a knife were to pass by one of these cognitive radar or magnetic sensors (or the combination of both working together), it would immediately send an alert to security officers, who could then identify and intercept the individual' (Griffiths 2020). Whilst the widespread adoption of this technology may provide the Metropolitan police with methods to identify individuals carrying knives, it is important to explore surveillance methods to pre-emptively identify individuals who are likely to be carrying before leaving their homes.

Mass surveillance technologies are not limited to physical observations (i.e. CCTV) but extend into the technical world. Since 2001 UK police forces have the ability to utilise the internet for investigative purposes with the creation of the National Hi-Tech Crime Unit (NHTCU) (Egawhary 2019). A growing area of interest has become the data rich world of social media. It has been said that social media platforms cannot only assist police with appealing information but also provide a platform to assist in intelligence-gathering (PoliceFoundation 2014). Whilst many predictions can be made from analysing social media data (Schoen et al. 2013), we ask the question of how effective social media surveillance and data collection can be for predicting crime or more specifically knife crimes. A study conducted by Wang et al. (2019) explored how analysing social media posts on drug related tweets over one year had a strong correlation to the crime rates of the same year. The study concludes that their findings show a strong possibility that police forces can utilise social media data to help predict crime hot-spots. Other research has also displayed the potential social media holds in analysing and predicting crime hot-spots. By analysing tweet through conducting sentiment analysis Mahajan and Mansotra state there are strong correlations between crime patterns predicted using tweets and the actual crime report statistics (Mahajan and Mansotra 2021). Whilst many papers highlight the uses of social media analysis for predictive policing (Aghababaei and Makrehchi 2016; Malleson and Andresen 2015), it is important to consider the ethical and moral implications of social media surveillance. Deliberation into the deployment of utilising social media data in policing is an important debate to ensure the ethical and legal boundaries are not crossed. A growing area of research in America has shown how the use of social media in pre-emptive policing has been met with questions of 'dragnet policing, harassment, and racial and religious profiling' (Patton et al. 2017). However, in the context of addressing knife-related crimes in London social media surveillance and analysis may be advantageous. With knife crime related incidents in London largely affiliated with organised gangs (Haylock et al. 2020) and reports showing the exploitation of social media for gang related activity (Whittaker et al. 2020), social media surveillance and analysis may be a turning point for predictive policing to tackle knife crime in London.

While we follow the definition of mass surveillance as 'Any monitoring, tracking, and otherwise processing of personal data of individuals in an indiscriminate or general manner, or of groups, that is not performed in a specific and lawfully 'targeted' way against a specific individual' (Montag et al. 2021, p. 10), as we stated it is important to consider the wider ethical concerns. As with anything there are two sides to consider when exploring mass surveillance in the context of policing. In some instances surveillance is a credible method to collect data and intelligence on a person of interest, whilst other case may not be justifiable. However, the ethics of surveillance is more complicated, there should be credible intelligence and reason for mass surveillance with the end goal of ultimately saving lives (Macnish 2017).The authors do not dispute the benefits of mass surveillance to tackle criminal activity especially knife-related crimes that may resolve in life-threatening instances, but the methods and justification must be sufficient to the cause.

6 The Use of Big Data to Predict Knife-Enabled Homicides or Knife Offences in the London Metropolitan Area

Broadly defined as the containment of large volumes of data, big data refers to the mass collection and analysis of data produced by companies and individuals that are inherently of large volume and require sophisticated storage and analysis considerations (Riahi and Riahi 2018). Moreover, big data is a term generated to keep up with the ever needing demand of businesses collecting and handling extreme data sets (Henke et al. 2016). Big data is generally described in relation to the 5 V's, Value, Veracity, Volume Variety and Velocity (Hadi et al. 2015). From scientific to medical, big data provided companies with copious benefits (Cole et al. 2015). Whilst in many industries the application of big data analytics has revolutionised operations, the implementation into policing has been a prolonged approached (Babuta 2017). With budget cuts looming over police forces (Dodd 2020) and the increase pressure of violent crime rates rising (Mcgleenon 2022), police forces in the UK must find innovative ways to make effective use of their data. A growing field of interest entitled 'data-driven policing' has began exploring how police forces can adopt data-driven technologies to keep up with the ever advancing technological evolution. Data-driven policing has been defined as 'the acquisition, analysis and use of a wide variety of digitised data sources to inform decision making, improve processes, and increase actionable intelligence for all personnel within a police service, whether they be operating at the front-line or in positions of strategic leadership' (Kearns and Muir 2019, p. 2). Besides allowing police forces to further analyse trends and patterns within data sets (Kaufmann et al. 2019), data-driven policing allows forces to be able to more efficiently deploy their scare resource (Hardy 2010). Whilst the authors do not dispute the benefits of intuitive policing and trusting gut instinct, utilising big data technologies for data-driven decision making provides credible insights (i.e. correlations and associations) and pattern/trend detection in data (Jeble 2018). Moreover, allowing forces to utilise technology to further enhance criminal investigations. Other uses of big data analytics include the possibility for forces to utilise pre-existing data to identify individuals of interest (Babuta 2017), ultimately addressing crime prevention, crime detection and national security (UK-Parliament 2014). However, it is important to consider that the quality of data-driven decision is largely dependent on the quality of data collected (Diván 2020). Additionally, it is vital to consider the perception and acceptance of these technologies when implementing rapidly into the public sector, for instance, making use of the Technology Acceptance Model (TAM) (Ariel 2019). This is especially important when the use of big data analytics creates huge change in police forces to overall traditional operations from reactionary policing (reacting to crime after it has occurred) to pre-emptive policing tactics (Hardyns and Rummens 2018). The authors are particularly interested in exploring how the benefits of data-driven policing (Big data analytics) and other forms of predictive analytics can help police in the London Metropolitan force navigate knife-related crimes. In order to correctly classify knife-related crimes for analysis, it is important to distinguish what is classified as a 'knife'. Pre-legislation classified knife crime inclusive of offensive weapons such as broken bottles. However, the Offensive Weapons Act of 2019 now states bladed and sharply pointed weapons are to be classified as a bladed article (Parliament: House of Commons 2019). Prior to 2018 knife crime statistics reported for England and Wales were the combination of knife and/or sharp instruments (Allen et al. 2018).

Interestingly, a report by Massey et al. found that 'Predicting which local areas are most likely to suffer knife-enabled homicides, based only on recent nonfatal knife injuries, can pinpoint risks of homicide in local areas that are up to 1400% higher than in most local areas, offering a range of strategies for resource allocation' (Massey 2019). Through analysing the data recorded from previous knife-enabled (nonfatal) crimes, the authors were able to gauge a better impression of fatal knife-enabled crimes the following year. Furthermore providing evidence of the potential for metropolitan police forces to forecast potential knife-related 'hot-spots'. The findings from the paper display a pragmatic approach to dealing with categorising knife crime. However, often when visualising knife crime related incidents in the London Metropolitan area, we are presented with the statistical outputs and findings for local authorities/boroughs. Moreover, data is accumulated for particular boroughs which may cover over 60 square miles. In order to accurately represent knife crime related incidents we must be able to visualise data on a Lower Layer Super Output Areas (LLSOA) (or even Output Areas) to directly pinpoint (to a street) knife crime hot-spots. As reinforced by Massey (2019), the current deployment of policing patrols are deployed across boroughs, but evidence suggests precise targeting data is needed to monitor hot-spots.

It is important to note that police forces not only have accesses to previous records but also can be seen as a data rich organisation who are constantly capturing real time information to be exploited (Ellison et al. 2021). Besides the typical analysis of big data to uncover trends and patterns, there is a growing interest in the area of artificial intelligence for proactive policing. The benefits of predictive software hold huge potential to being able to identify future offenders and predict locations of crimes; however, it is important to consider the wider implications of such system and the accountability and operational transparency required (Oatley 2021). An interesting approach to tackling knife crime through big data analytics has been explored by West Midlands police. Through following the same approaches of exploratory (Spatial) data analysis, West Midland police combined their analysis with additional statistics of 2019 knife crime data. These included the distribution of monthly counts, knife crime by day, knife crime by hours, time between sequential knife crimes and spatial distance between crime (WestMidlandPCC 2021). This allowed West Midlands police to conduct an enhanced analysis of 2019 data and encode further dimensions into their LLOSA visualisations.

Despite the current adoption of big data analytics in policing, there are still many areas where further improvement of the extraction of information and knowledge can be used to decrease crime rates (Ridgeway 2018). However, detailed in the Association of Police and Crimes Commissioners report on the UK policing vision for 2025 it highlights the acknowledgment that policing must continue to embrace new technology for the opportunities it presents in twenty-first century policing (NPCC 2025).

6.1 How Can Visualisations Aid Pre-emptive Policing?

Visualisations offer a visual depiction of data in a graphical format to transform raw data into meaningful insights. Moreover, a visualisation has the ability to accelerate a person's perception of the data by providing crucial insights and improving decision making (Al-Kassab et al. 2014). With humans collecting the most information through visuals than any other sense, the visual depiction of data can provide the highest bandwidth for the communication of data between computer and human (Ware 2004). In the Metropolitan police force, visualisations offer a unique ability to be able to view the masses amounts of data collected in an easily interpreted format. In general law enforcement visualisations offer the ability to conduct both crime pattern analysis and criminal association discovery. Crime pattern analysis utilises spatial data of criminal events to help interrupt patterns of crimes, whilst criminal association discovery utilises various sources of data (i.e. police records) to discover associations and build leads (Chen et al. 2005). In the Metropolitan Police force data visualisations allow for officers to understand the relationship between knife-related crimes and perpetrators (Frater and Gamman 2020). Whilst British police forces have access to state-of-the-art visualisations, the British public also have access to track and visualise spatial knife crime data using Google maps. The Google maps knife crime tracker provides users with a map of the UK with pinpoints of knife crime incidents, the data of occurrence and details of the incident (Clouhhd 2022).

7 Conclusion

Through analysing contemporary literature this chapter has informed on the ever-growing knife crime epidemic specifically in the London boroughs. Through reporting on the current statistics surrounding knife crime, the chapter displays the severity of knife-related incidents in the London area. The chapter critically analyses the key reasons as to why knife crime is on an upwards trends through documenting the motivations and risks associated with carrying a bladed article. The analysis of the knife crime epidemic within London has shown the ever changing policing strategies to maintain peace and order. Forces are dynamically adapting to follow new procedures such as enhanced stop and searches to help battle the widespread issues. However, this chapter reports on how cutting edge technology can aid the London Metropolitan police to identify individuals who are carrying or likely to carry a knife. Whilst traditional policing has proven effective in the past, new pre-emptive policing must embrace new technologies to reap the benefits. Evidence in other sectors has shown the benefits that technology and specifically big data can bring for a competitive edge. The authors feel with the right adoption and widespread implementation, the use of big data analytics can help maximise allocating officers to strategically placed locations. However, it is important to note that the use of innovative technologies is not proposed as a means to reduce the number of officers. Instead, technology should be used as a means to help officers make decisions strategically and optimise their time (Griffiths 2020).

Acknowledgments Supported by Knowledge Economy Skills Scholarships 2 (KESS2) which is an All Wales higher-level skills initiative led by Bangor University on behalf of the HE sectors in Wales. It is part-funded by the Welsh Government's European Social Fund (ESF) competitiveness programme for East Wales.

References

Aghababaei, S., & Makrehchi, M. (2016). Mining social media content for crime prediction. In *IEEE/WIC/ACM International Conference on Web Intelligence (WI)* (pp. 526–531). IEEE Xplore. https://doi.org/10.1109/WI.2016.0089.

Al-Kassab, J., Ouertani, Z., Schiuma, G., & Neely, A. (2014). Information visualisation to support management decisions. *International Journal of Information Technology and Decision Making, 13*(2), 407–428.

Allen, G., Audickas, L., Loft, P., & Bellis, A. (2018). Knife crime in England and Wales – Briefing paper – SN4304 (pp. 1–27). https://www.basw.co.uk/system/files/resources/SN04304.pdf

Allen, G., Audickas, L., Loft, P., & Bellis, A. (2021). Knife crime in England and Wales – house of commons library. In *Commons Research Briefing (SN4304)* (pp. 1–29). https://commonslibrary.parliament.uk/research-briefings/sn04304/

Allen, G., Lukas Audickas, P., & Loft, A. B. (2019). Knife crime in England and Wales. Technical report. House of Commons Library. https://researchbriefings.parliament.uk/ResearchBriefing/Summary/SN04304.

Ariel, B. (2019). *Advocate: Technology in policing advocate technology in policing*. Cambridge: Cambridge University Press. https://doi.org/10.1017/9781108278423.022.

Ashby, M. P. J. (2017). The value of CCTV surveillance cameras as an investigative tool: An empirical analysis. *European Journal on Criminal Policy and Research, 23*(3), 441–459. https://doi.org/10.1007/s10610-017-9341-6.

Babuta, A. (2017). Big data and policing. Technical report. Royal United Services Institute.

Barajas-Gonzalez, R. G., Ayón, C., Brabeck, K., Rojas-Flores, L., & Valdez, C. R. (2021). An ecological expansion of the adverse childhood experiences (ACEs) framework to include threat and deprivation associated with U.S. immigration policies and enforcement practices: An examination of the Latinx immigrant experience. *Social Science and Medicine, 282*. https://doi.org/10.1016/j.socscimed.2021.114126.

Brennan, I. (2018). Knife crime: Important new findings could help us understand why people carry weapons. The Conversation. https://theconversation.com/knifecrimeimportantnewfindingscouldhelpusunderstandwhypeoplecarryweapons.

Brown, E. L., Ware, G., & Cassimally, K. A. (2019) Knife crime: causes and solutions – editors' guide to what our academic experts say. The Conversation. http://theconversation.com/knifecrimecausesandsolutionseditorsguidetowhatouracademicexpertssay113318.

Buckchash, H., & Raman, B. (2017). A robust object detector: application to detection of visual knives. In *IEEE International Conference on Multimedia and Expo Workshops (July)* (pp. 633–638).

Chakelian, A. (2019). Cocaine clichés and London blinkers: yet another knife crime myth-buster. https://www.newstatesman.com/politics/uk/2019/03/cocaine-clich-s-and-london-blinkers-yet-another-knife-crime-myth-buster.

Chen, H., Atabakhsh, H., Tseng, C., Kaza, S., Eggers, S., Gowda, H., Petersen, T., Violette, C., & Ave, S. S. (2005). Visualization in law enforcement (May 2014). https://doi.org/10.1145/1056808.1056893.

Clouhhd, G. M. (2022). Knife crime tracker. Tracking Stabbings across the UK.

Cole, D., Nelson, J., & Mcdaniel, B. (2015). Benefits and risks of big data. In *SAIS 2015 Proceedings* (Vol. 26, pp. 1–6). SAIS.

CPS. (2021). Offensive weapons, knife crime practical guidance. Technical report. https://www.cps.gov.uk/legal-guidance/offensive-weapons-knife-crime-practical-guidance.

Davenport, J. (2022). Met may start trialling thermal-image scanners for knife searches. Technical report. Evening Standard, London. https://www.standard.co.uk/news/crime/.

Deloitte. (2015). The Digital Policing Journey: From Concept to Reality Realising the benefits of transformative technology. Technical report. Deloitte, United Kingdom. https://www2.deloitte.com/content/dam/Deloitte/uk/Documents/public-sector/deloitte-uk-ps-digital-police-force.pdf.

Diván, M. J. (2020). Data-driven decision making. In *2017 International Conference on Infocom Technologies and Unmanned Systems (Trends and Future Directions) (ICTUS)* (pp. 50–56). No. December 2017. IEEE Xplore. https://doi.org/10.1109/ICTUS.2017.8285973.

Dodd, V. (2020). Police in England and Wales facing 'new era of austerity. https://www.theguardian.com/uk-news/2020/jul/01/police-warn-of-cuts-to-funding-even-worse-than-in-austerity-years.

Egawhary, E. M. (2019). Article the surveillance dimensions of the use of social media by UK police forces. *Platform Surveillance, 17*(1), 89–104.

Ellison, M., Bannister, J., & Haleem, M. S. (2021). Understanding policing demand and deployment through the lens of the city and with the application of big data. *Urban Studies, 58*(15), 3157–3175. https://doi.org/10.1177/0042098020981007.

Ford, R. G. (2018). Matt: Young people, violence and knives – revisiting the evidence and policy discussions. Centre for Crime and Justice Studies The Hadley Trust. https://www.crimeandjustice.org.uk/sites/crimeandjustice.org.uk/files/Knifecrime.November.pdf.

Frater, A., & Gamman, L. (2020). Beyond knife crime: Towards a design-led approach to reducing youth violence. RYF (pp. 1–99). https://www.arts.ac.uk/__data/assets/pdf_file/0028/249346/Beyond-Knife-Crime-Report.pdf

Galab, M. K., Taha, A., & Zayed, H. H. (2021). Adaptive technique for brightness enhancement of automated knife detection in surveillance video with deep learning. *Arabian Journal for Science and Engineering, 46*(4), 4049–4058. https://doi.org/10.1007/s13369-021-05401-4.

Griffiths, S. (2020). No TitleTech tackles violent crime: Knife crime across the UK has risen steeply in recent years – can advanced technology help the police put an end to such violence? *Engineering & Technology, 15*(3), 22–25. https://doi.org/10.1049/et.2020.0300.

Hadi, H. J., Shnain, A. H., Hadishaheed, S., & Ahmad, A. H. (2015). Big data and five V's characteristics. *International Journal of Advances in Electronics and Computer Science, 2*(1), 16–23.

Hardy, E. (2010). Geography public safety. Technical report 3. Office of Safety Programs.

Hardyns, W., & Rummens, A. (2018). Predictive policing as a new tool for law enforcement? Recent developments and challenges. *European Journal on Criminal Policy and Research, 24*(3), 201–218. https://doi.org/10.1007/s10610-017-9361-2.

Haylock, S., Boshari, T., Alexander, E. C., Kumar, A., Manikam, L., & Pinder, R. (2020). Risk factors associated with knife-crime in United Kingdom among young people aged 10–24years: A systematic review. *BMC Public Health, 20*(1), 1451. https://doi.org/10.1186/s12889-020-09498-4.

Henke, N., Bughin, J., Chui, M., Manyika, J., Saleh, T., Wiseman, B., & Sethupathy, G. (2016). The age of analytics: Competing in a data-driven world McKinsey & Company (December). http://www.mckinsey.com/business-functions/mckinsey-analytics/our-insights/the-age-of-analytics-competing-in-a-data-driven-world.

House of Lords. (2019). Knife crime: policy and causes. Technical report. https://researchbriefings.files.parliament.uk/documents/LLN-2019-0061/LLN-2019-0061.pdf.

Houses of Parliament. (2019). Early interventions to reduce violent crime. Technical report. https://researchbriefings.files.parliament.uk/documents/POST-PN-0599/POST-PN-0599.pdf.

Jeble, S. (2018). Role of big data in decision making. *OSCM Publications, 11*(1). https://doi.org/10.31387/oscm0300198.

Kaufmann, M., Egbert, S., & Leese, M. (2019). Predictive policing and the politics of patterns. *The British Journal of Criminology, 59*, 674–692. https://doi.org/10.1093/bjc/azy060.

Kearns, I. A. N., & Muir, R. (2019). Data-drive policing and public value. Technical report. March, The Police Foundation. https://www.police-foundation.org.uk/2017/wp-content/uploads/2010/10/data_driven_policing_final.pdf.

Lemos, G. (2004). Fear and fashion The use of knives and other weapons by young people. https://lemosandcrane.co.uk/resources/L&C-FearandFashion.pdf.

Macnish, K. (2017). *The ethics of surveillance* (1 edn.). London: Routledge.

Mahajan, R., & Mansotra, V. (2021). Correlating crime and social media: Using semantic sentiment analysis. *International Journal of Advanced Computer Science and Applications, 12*(3), 309–316.

Malleson, N., & Andresen, M. (2015). The impact of using social media data in crime rate calculations: Shifting hot spots and changing spatial patterns. *Cartography and Geographic Information Science, 42*(2), 112–121.

Marshall, B., Webb, B., & Tilley, N. (2005). Rationalisation of current research on guns, gangs and other weapons: Phase 1 http://nomsintranet.org.uk/roh/official-documents/Marshall_Webb_Tilly_gangs_and_guns_2005.pdf.

Massey, J. (2019). Forecasting knife homicide risk from prior knife assaults in 4835 local areas of London, 2016–2018. *Cambridge Journal of Evidence-Based Policing, 1*(3), 1–20.

Mayor of London. (2021). Police and crime plan 2021–2025. https://www.london.gov.uk/publications/police-and-crime-plan-2021-25.

McCallum, A. (2020). Digital policing strategy. Technical report. One Met, London. https://www.met.police.uk/SysSiteAssets/media/downloads/force-content/met/about-us/one-met-digital-policing-strategy-2017-2020.pdf.

Mcgleenon, B. (2022). Violent crime rises as 'feral youths' stalk streets of British city leaving 'no-go areas'. https://www.express.co.uk/news/uk/1568647/Violent-crimecanterburyferalyouthBritishcityattackantisocial-behaviour.

Met Police Data Board. (2021). MPS FY 2020/21 crime statistics. Technical report. https://www.met.police.uk/sd/stats-and-data/met/year-end-crime-statistics-20-21/.

Metropolitan Police. (2022). The structure of the Met. https://www.met.police.uk/police-forces/metropolitan-police/areas/about-us/about-the-met/structure/.

Metropolitan Police Service. (2021). Met strategic digital enabling framework 2021–25. Technical Report. https://www.met.police.uk/SysSiteAssets/media/downloads/force-content/met/about-us/met-strategic-digital-enabling-framework-2021-2025.pdf.

Montag, B. L., Mcleod, R., Mets, L. D., Gauld, M., Rodger, F., & Pe, M. (2021). A legal analysis of bio-metric mass. Technical report. European Digital Rights, Europe.

National Crime Agency. (2019). County lines drug supply, vulnerability and harm 2018. Technical report. https://www.nationalcrimeagency.gov.uk/who-we-are/publications/257-county-lines-drug-supply-vulnerability-and-harm-2018/file.

NPCC. (2025). Policing Vision 2025. https://www.npcc.police.uk/documents/PolicingVision.pdf.

Oatley, G. C. (2021). Themes in data mining, big data, and crime analytics. In W. Pedrycz (ed.), *Wires data mining and knowledge discovery* (pp. 1–19). No. August. https://doi.org/10.1002/widm.1432.

Parliament: House of Commons.: Offensive weapons act 2019 c.17. (2019). https://www.legislation.gov.uk/ukpga/2019/17/contents/enacted.

PatriotOne. (2022). Threat detection solutions. https://patriot1tech.com/.

Patton, D. U., Brunton, D. W., Dixon, A., Miller, R. J., Leonard, P., & Hackman, R. (2017). Stop and frisk online: theorizing everyday racism in digital policing in the use of social media for identification of criminal conduct and associations. *Social Media + Society, 3*(3), 1–10. https://doi.org/10.1177/2056305117733344.

PoliceFoundation. (2014). Police use of social media. Technical report. PoliceFoundation, United Kingdom.

Policy Exchange. (2021). Knife crime in the capital. https://policyexchange.org.uk/wp-content/uploads/Knife-Crime-in-the-Capital.pdf.

Riahi, Y., & Riahi, S. (2018). Big data and big data analytics: Concepts, types and technologies big data and big data analytics: Concepts, types and technologies (November). https://doi.org/10.21276/ijre.2018.5.9.5.

Ridgeway, G. (2018). Policing in the era of big data. *Annual Review of Criminology, 1*(1), 401–419. https://doi.org/10.1146/annurev-criminol-062217-114209.

Schoen, H., Gayo-avello, D., Metaxas, P., & Mustafaraj, E. (2013). The power of prediction with social media. *Internet Research: Electronic Networking Applications and Policy, 23*, 1–21. https://doi.org/10.1108/IntR-06-2013-0115, https://www.police-foundation.org.uk/2017/wp-content/uploads/2017/08/Social_media_briefing_FINAL.pdf.

Sheldon, B. (2012). Camera surveillance within the UK: Enhancing public safety or a social threat? *International Review of Law, Computers & Technology, 25*:3, 193–203. https://doi.org/10.1080/13600869.2011.617494

Silvestri, A., Oldfield, M., Squires, P., & Grimshaw, R. (2009). Young people, knives and guns a comprehensive review, analysis and critique of gun and knife crime strategies. Centre for Crime and Justice Studies. https://www.crimeandjustice.org.uk/sites/crimeandjustice.org.uk/files/YPknivesandguns.pdf.

Sintonen, L., Turtiainen, H., Costin, A., Hamalainen, T., & Lahtinen, T. (2021). OSRM-CCTV: Open-source CCTV-aware routing and navigation system for privacy, anonymity and safety. Preprint. CoRR abs/2108.09369.

Timberlake, A. (2020). Using new technologies and data to combat knife crime. *The AI Journal*. https://aijourn.com/using-new-technologies-and-data-to-combat-knife-crime/.

UK-Parliament. (2014). Big data, crime and security. Technical Report 470, Parliamentary office of Science and Technology, United Kingdom.

Wang, Y., Yu, W., Liu, S., & Young, S. D. (2019). The relationship between social media data and crime rates in the United States. *Social Media + Society*. https://doi.org/10.1177/2056305119834585.

Ware, C. (2004). *Information visualization: perception for design* (3 edn.). Burlington: Morgan Kaufmann.

Warrell, H. (2018). Why England is facing a rising tide of knife crime. https://www.ft.com/content/0a4166b8-e8ef-11e8-a34c-663b3f553b35.

Webster, W. (2009). CCTV policy in the UK: Reconsidering the evidence base. *Surveillance and Society* ***6***, 10–22. https://doi.org/10.24908/ss.v6i1.3400.

WestMidlandPCC. (2021). Knife crime prediction. Technical Report. December 2020, West Midlands Police. https://www.westmidlands-pcc.gov.uk/wp-content/uploads/2021/10/2021-07-21-EC-Agenda-Item-4-Short-Term-Knife-Predictions.pdf?x49252.

Wheller, A. M. (2019). Levin: Knife crime evidence briefing. College of Policing. https://whatworks.college.police.uk/Research/Documents/Knife_Crime_Evidence_Briefing.pdf.

Whittaker, A., Densley, J., & Moser, K. S. (2020). No two gangs are alike: The digital divide in street gangs' differential adaptations to social media Computers in Human Behavior No two gangs are alike: The digital divide in street gangs' differential adaptations to social media. *Computers in Human Behavior, 110*, 1–10 . https://doi.org/10.1016/j.chb.2020.106403.

Williams, C. A. (2003). Police surveillance and the emergence of CCTV in the 1960s. *Crime Prevention and Community Safety, 5*, 27–37. https://doi.org/10.1057/palgrave.cpcs.8140153.

Printed in the United States
by Baker & Taylor Publisher Services